ISBN 978-1-332-33969-3
PIBN 10316204

WOODEN BOX

AND

CRATE CONSTRUCTION

PREPARED BY

FOREST PRODUCTS LABORATORY

U. S. DEPARTMENT OF AGRICULTURE

FOREST SERVICE

MADISON, WISCONSIN

PUBLISHED BY

NATIONAL ASSOCIATION OF BOX MANUFACTURERS

1553 CONWAY BUILDING

CHICAGO, ILLINOIS

1921

PUBLISHER'S INTRODUCTION

The National Association of Box Manufacturers is an organization of the leading manufacturers' of wooden boxes of every section of the country. Many of its members also make other types or kinds of boxes, but the association confines its activities to the interests of the wooden box.

The association aims to promote all projects of genuine interest to its members; to obtain a unity of action in matters affecting the wooden box industry; to advertise properly the merits of wooden boxes, and to be, in fact, a constructive influence in the improvement of conditions in the industry and better service for consumers. And by the same token the association stands guard against any project or practice which may, in any way, bring wooden box service into disfavor or disrepute.

The association believes in the scientific construction of packing boxes. It believes that the dissemination of knowledge among box users, as well as manufacturers, as to what has already been accomplished and what is now being done along the lines of scientific research in box construction, is to the mutual advantage of all. Such knowledge enlarges the field of usefulness of the wooden box, and leads to conservation of material and lowering of costs, all of which tends to stabilize the use of wooden boxes, and the business of the manufacturer. This is the reason for publishing this book.

The book includes, in a general way, all the latest and most accurate information available on box and crate design. However, details applicable to different kinds of boxes for carrying special commodities vary with each commodity, hence the application of the fundamental principles discussed in this book must be made by each manufacturer and individual user of such boxes, except, of course, as the principles apply to boxes which have already been standardized as to specifications.

Standardization of boxes and crates is a recognized aim of the association. Many packages have already been standardized and many others are being studied with a view to standardization.

458422

Much valuable information on boxes for carrying certain commodities has been obtained as a result of these studies. No attempt has been made to include that information in this book, but it is available and will be freely given to those who have need for it.

All of the machinery of the association is at the service of the public in any matters pertaining to the development of better packing methods or in solving any packing problem.

THE NATIONAL ASSOCIATION
OF BOX MANUFACTURERS.

1553 Conway Building.
Chicago, Illinois,
March 1, 1921.

INTRODUCTION

The Forest Products Laboratory is a research unit of the Forest Service, U. S. Department of Agriculture. It is located at Madison, Wisconsin, where it was established in 1910 in co-operation with the University of Wisconsin. Its purpose is to acquire, disseminate and apply useful knowledge of the properties, uses and methods of utilization of all forest products and thereby to promote economy and efficiency in the processes by which forests are converted into commercial products. In this work 220 research technologists are employed. Its field of investigations and activities embraces:

1. Obtaining authoritative information on the mechanical and physical properties of commercial woods and products derived from them;

2. Studying and developing the fundamental principles underlying the seasoning and kiln-drying of wood, its preservative treatment, its use for the production of· fiber products (pulp, paper, fiber board, etc.), and its use in the manufacture of alcohol, turpentine, rosin, tar and other chemically derived products;

3. Developing practical ways and means of using wood which, under present conditions, is being wasted;

4. Co-operating with consumers of forest products in improving present methods of use; also in formulating specifications and grading rules for commercial woods and materials obtained from them, and for material used in the preservative treatment of wood; and

5. Making the information obtained available to the public through publications, correspondence, and by other means.

Commercial research and mechanical tests on containers were begun in 1915 in co-operation with the National Association of Box Manufacturers, the National Canners' Association, and the National Wholesale Grocers' Association. Methods of testing and testing equipment were developed which have since become more or less standard for the box industry. From the data accumulated by the laboratory, the War Department prepared during the war general specifications for overseas containers. At the laboratory in Madison, many containers to be used for the shipment of war equipment were tested and redesigned along more economical lines, including increase in strength, decrease in amount of material required, decrease in cubic contents and a consequent reduction of ocean freight costs.

Since the war the laboratory has co-operated with many associations and companies in testing and studying the construction of many different types of containers such as those used for the shipment of electric lamps, cream separators, small tractors, talking machines, boiler castings, furniture, paints and oils, piano benches, fruit baskets and crates, and shoes.

CONTENTS

CHAPTER I

Use of Wood in Box and Crate Construction

ix

CONTENTS

CHAPTER II

Box Design

CHAPTER III

Crate Design

CHAPTER IV

Box and Crate Testing

CHAPTER V

Box and Crate Specifications

CHAPTER VI

Structure and Identification of Woods

TABLES

PLATES

FIGURES

WOODEN BOX AND CRATE CONSTRUCTION

CHAPTER I

THE USE OF WOOD IN BOX AND CRATE CONSTRUCTION

COMMERCIAL GRADES AND SIZES OF LUMBER AVAILABLE FOR BOX CONSTRUCTION—IMPORTANT PHYSICAL PROPERTIES OF WOOD WHICH INFLUENCE ITS USE IN BOX CONSTRUCTION —MECHANICAL OR STRENGTH PROPERTIES OF WOOD—CARE AND SEASONING OF LUMBER IN STORAGE—THE USE OF VENEER IN THE CONSTRUCTION OF PACKING BOXES.

ADAPTABILITY

Availability—Although our forest resources are rapidly diminishing, lumber is still the most abundant material economically suitable for the manufacture of packing boxes and crates. The lower grades of lumber are used almost exclusively for this purpose and are sure to be abundant for many years to come. The closer utilization of virgin timber, especially of the small and defective trees and of the tops, and the cutting of second and third-growth timber, which is nearly always of poorer quality, make a large amount of low grade lumber available. Thus, in the eastern section of the United States, cut-over or second-growth forests furnish the greater part of the supply of logs, and these logs furnish a large amount of low grade lumber, so that the percentage of low grades manufactured from southern yellow pine, hemlock, and northern pine has increased rapidly during recent years.

Cost—The large amount of low-grade lumber manufactured has kept the cost comparatively low. And the increase in cost of rotary-cut lumber or veneer, caused by the necessity of using better logs than for other kinds of box lumber, is offset by the thinness of the material and the comparatively small waste in cutting.

Salvage Value—Wooden packing boxes have considerable salvage value, as is indicated by the ready sale merchants have for such material. Some large mercantile concerns rebuild their used boxes to store goods in or to use for hipping purposes. Whenever boxes are used over again for shipping, special precautions should be taken to make sure that they are in fit condition for transportation. Many of the claims for damaged freight are due to the use of second-hand boxes in imperfect condition. Other purposes to which used packing boxes are put are commonplace. Quite often the box is taken apart and the lumber used for repairs or light construction work. Considerable kindling is often obtained from the dump pile of boxes in the merchant's back yard.

Qualities

Desirable Qualities—Wood possesses the following properties desirable in the manufacture and subsequent use of packing boxes and crates:

1. It is strong for its weight.
2. It is easily worked.
3. It is easily fastened together.
4. It is not easily indented or bent out of shape
5. It is not corroded by sea water or attacked by acids unless they are of high concentration.
6. It is a poor conductor of heat, which is of service in protecting certain contents of packing boxes when stored for a relatively short time exposed to the sun's rays, to heat from steam pipes or boilers, or to freezing temperatures.
7. It takes and holds ink well, so that addresses and advertising can be stenciled or printed on it without difficulty.
8. When its usefulness is ended and it becomes refuse, it can be easily disposed of

KINDS AND AMOUNTS OF LUMBER USED

The Choice of Species—The choice of species for boxes and crates is influenced chiefly by cost and availability, the tendency being usually to make selection out of kinds of lumber that are comparatively cheap and readily available in the locality where the manufacturing plant is situated; but, in addition, the degree in which the desirable qualities listed above are present is also considered, together with resistance to shearing out of fastenings at the end of boards, the commodities to be packed, and the comparative freedom from odor and taste.

TABLE 1. Box Woods, Consumption by Box Manufacturers, and Total
Lumber Production

Kind of wood	Quantity used annually by box manufacturers (1912)	Total lumber production[1] (1918)
	Feet board measure	Feet board measure
White pine	1,131,969,940	2,200,000,000
Yellow pine (including North Carolina pine)	1,042,936,123	10,845,000,000
Red gum (including sap gum)	401,735,390	765,000,000
Spruce	335,935,643	1,125,000,000
Western yellow pine	288,691,927	1,710,000,000
Cottonwood	210,819,509	175,000,000
Hemlock	203,526,091	1,875,000,000
Yellow poplar	165,116,737	290,000,000
Maple	96,831,648	815,000,000
Birch	90,787,900	370,000,000
Basswood	86,979,611	200,000,000
Beech	77,899,280	290,000,000
Tupelo	74,982,910	237,000,000
Elm	63,726,458	195,000,000
Oak	56,362,111	2,025,000,000
Balsam fir	40,173,700	82,000,000
Cypress	38,962,895	630,000,000
Chestnut	36,216,700	400,000,000
Sugar pine	24,686,000	111,800,000
Sycamore	16,451,693	30,000,000
Ash	10,507,308	170,000,000
Willow	10,004,600	6,269,000
Larch (including tamarack)	7,470,300	355,000,000
Douglas fir	7,349,840	5,820,000,000
Noble fir	6,653,500	5,201,000
Magnolia	5,449,000	1,579,000
Buckeye	3,174,028	3,646,000
White fir	3,142,080	213,000,000
Cedar	2,512,150	245,000,000
Redwood	2,439,500	443,231,000
Red fir	1,328,330	Included in white fir
All other woods	3,150,278	60,963,000
Total	4,547,973,180	31,694,689,000

Amount of Each Kind of Lumber Used[2]—The woods
used for boxes, the amounts, and the total annual production
of lumber of each species are given in Table 1. The figures
on the consumption of box lumber are those secured by a

---

[1]Computed total productions. For details see U. S. Department of Agriculture
Bulletin 845, "Production of Lumber, Lath, and Shingles in 1918."
[2]The explanation of Tables 1, 2 and 3, and the information contained in Table
5 were taken principally from a mimeographed circular. "Packing Box Woods:
Kinds, Supply, Distribution, Grades and Sizes Available," prepared by J. C. Nellis,
formerly Forest Examiner in the Forest Service, U. S. Dept. Agriculture.

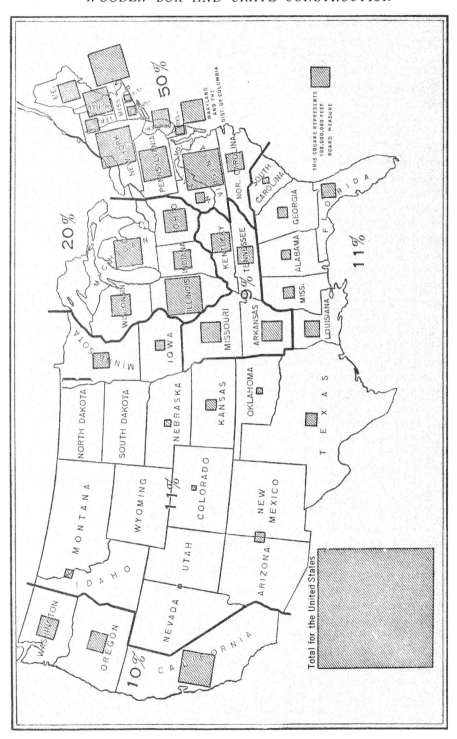

series of studies of the wood-using industries by States, made by the Forest Service from 1909 to 1913. The information, however, is not complete owing to the fact that a number of mills did not report their consumption. It is estimated that today the amount of lumber used annually in the manu-. facture of shipping containers is between five and six billion feet, or nearly one-sixth of the total lumber cut in the United States. While the figures in the table apply to different years, all are based on a period of 12 months. They may be said to apply to 1912 as an average year.

The figures on the total production of different species are for 1918, the latest available at this time.

In compiling statistics it has been found best to combine some of the different species or kinds of woods because of the confusion on the part of manufacturers as to species. Some of the names listed in Table 1 cover a number of species.

Consumption of Box Lumber by States—The total amount of wood consumed in each State by box manufac turers and the total lumber cut of each State are given in Table 2. This is shown graphically in figure 1, in which the size of the squares in each State represents the relative total consumption of wood for boxes, crates, and fruit and vegetable packages. The statistics used in Table 2 come from the same source as those in Table 1. Table 3 shows the principal woods used in the important box manufactur ing States. This table is limited to include only the 24 most important box manufacturing States and the 19 most im portant species; but the 24 States are located in all parts of the country, so that the table shows what woods are most used in every part of the United States.

Distribution of Box Lumber—Some of the largest box manufacturing States produce but little of the lumber so used. In this class are Illinois, Pennsylvania, and New York. Other States which are of medium rank in the box industry, but produce very little box lumber, are New Jersey, Maryland, Ohio, and Indiana.

From the region south of the Ohio and Potomac Rivers box lumber moves northward to consuming points in two general currents, separated by the Appalachian Mountains; that is, North Carolina pine from Virginia and the Caro-linas goes northward east of the Allegheny Mountains, while from the Central and Gulf States yellow pine, red gum, etc.,

TABLE 2. BOX LUMBER CONSUMPTION AND TOTAL LUMBER PRODUCTION
BY STATES

State	Quantity used annually for boxes (1912)	Total lumber production (1918)
	Feet board measure	Feet board measure
Virginia	433,028,997	855,000,000
New York	390,057,650	335,000,000
Illinois	389,199,000	42,000,000
Massachusetts	353,405,350	175,000,000
California	309,406,285	1,277,084,000[1]
Pennsylvania	276,587,094	530,000,000
Michigan	232,111,486	940,000,000
New Hampshire	200,209,596	350,000,000
Ohio	153,417,273	235,000,000
Maryland	144,309,000	71,000,000
Wisconsin	119,267,000	1,275,000,000
Kentucky	112,424,500	340,000,000
Missouri	111,765,699	273,000,000
Arkansas	110,822,000	1,470,000,000
Maine	108,889,400	650,000,000
New Jersey	102,605,205	19,500,000
Washington[2]	96,448,500	4,603,123,000
Indiana	85,267,160	250,000,000
Oregon[2]	78,939,000	2,710,250,000
Tennessee	77,979,510	- 630,000,000
Minnesota	77,854,600	1,005,000,000
North Carolina	76,525,000	1,240,000,000
Louisiana	56,004,500	3,450,000,000
Florida	53,469,000	950,000,000
Vermont	48,871,060	160,000,000
Mississippi	39,295,093	1,935,000,000
Texas	35,762,125	1,350,000,000
Iowa	31,340,476	14,200,000
Kansas	28,544,500	8,401,000
Arizona and New Mexico	28,035,000	172,576,000
Delaware	27,624,175	6,000,000
Connecticut	24,411,090	64,000,000
Georgia	24,373,409	515,000,000
West Virginia	23,837,000	720,000,000
Alabama	22,442,000	˙1,270,000,000
Rhode Island	15,951,200	13,100,000
South Carolina	13,960,000	545,000,000
Idaho	10,245,000	802,529,000
Nebraska	6,861,000	none
Montana	5,249,927	340,000,000
Colorado	4,734,000	56,882,000
Oklahoma	4,389,000	195,000,000
Nevada and Utah	1,517,000	9,815,000[3]
District of Columbia	518,655	none
North and South Dakota	18,667	29,533,000[4]
Wyoming	none	7,501,000
Total	4,547,973,180	31,890,494,000

[1]California and Nevada.
[2]1914 statistics on box lumber consumption are available: Washington, 106,-307,980 feet board measure; Oregon, 72,299,344 feet board measure.
[3]Utah only. [4]South Dakota only.

TABLE 3. QUANTITIES OF PRINCIPAL WOODS USED ANNUALLY BY BOX AND CRATE MANUFACTURERS IN THE MOST IMPORTANT STATES

(In thousands of feet B. M.)

State	Sugar pine	Chestnut	Cypress	Balsam fir	Oak	Elm	Tupelo	Beech	Basswood	Birch	Maple	Yellow poplar	Hemlock	Cottonwood	Western Yellow pine	Spruce	Red gum	Yellow pine (incl. N.C. pine)	White pine
Virginia		4,161	3,634		60	140	15,920		15			14,611	100			250	29,598	357,203	3,816
New York		4,832	5,524		3,765	9,123	251	11,381	11,716	2,550	6,335	9,666	11,167	13,231		50,316	3,455	108,248	137,577
Illinois		200	910		550	13,174	5,155	3,713	12,212	39,458	28,519	12,504	34,477	42,942		363	57,894	27,209	105,638
Massachusetts		2,362		100	281	879		310	605	347	729	6,723	27,394	22	225,029	31,972		2,442	263,443
California	20,536			14,879									500	2,217		46,621			
Pennsylvania		12,077	1,719		7,729	544	6,568	13,584	3,156	6,069	13,608	18,576	9,769	4,680		14,648	12,806	98,318	51,903
Michigan		340		418	698	15,244	200	28,514	19,140	11,953	29,217	342	29,022	6,443		733	5,005	10,123	66,553
New Hampshire																			146,711
Ohio		153	2,824	3,865	840	7,074	850	70	486	70	40	38,779	20,035	425		26,558	5,623	301	35,949
Maryland		6,343	3,320		6,199	200	8,990	7,338	4,333	765	8,007	3,813	1,267	10,540		1,758	2,985	13,580	4,195
Wisconsin		694		1,500		829		100	100	200	60	400	22	25			190	119,305	59,900
Kentucky			2,100		1,825	1,820		2,950	13,515	15,597	1,708	27,148	17,657	1,300			37,275	350	1,226
Missouri		116	5,533		607	2,483	711	600	2,857		1,505	145	2,000	12,734			69,601	10,595	505
Arkansas			2,910		17,500	1,528	3,100		177	200	304			17,518			48,538	12,139	
Maine				16,248						3,180	225		4,704	8,461		21,873		27,265	60,876
New Jersey		196	110		31	37	750		301		62	1,115	5,904	1,352		16,637	1,624	39,147	38,113
Washington (1914)									400						37,155	52,849			2,940
Indiana (1914)			2,590		6,443	7,131	1,900	5,345	2,097	4,675	1,613	4,930	9,614	500			16,451	15,916	7,965
Oregon (1914)	14,200				945	240	871	490	1,796	1,114	408	16,186	502	6,868	16,992	8,244	16,743	13,188	20,673
Tennessee		520				386			10,711		14		5,813	22,550					934
Minnesota				25										360					60,771
North Carolina			250			100	8,000		160		650	2,938	3,858			520	10,980	47,247	6,310
Louisiana		1,030	4,230		400		6,265					50		12,710			10,622	11,494	
Florida			49				33										739	49,830	

move approximately straight northward. Material from the Lake States, principally Minnesota, Wisconsin, and northern Michigan, moves to northern Illinois and Indiana. Canadian white pine goes into Michigan, northern Ohio, and western New York. New England is to a considerable extent self-supporting as regards box lumber, although considerable North Carolina pine gets into Connecticut and Rhode Island and some Canadian white pine goes into northern New England. The Western Pacific States are not only self supporting but at times supply considerable material to the territory east of them. Some box material is shipped from Arizona and New Mexico, principally to the middle west. Mexican pine is used by some of the box factories on the Mexican border.

As the outcome of conditions resulting from the war, many of the different species may now be found in markets far removed from their regions of growth, e. g., white pine from New England is manufactured into boxes in Texas and spruce from Oregon and Washington in Massachusetts.

COMMERCIAL GRADES AND SIZES OF LUMBER AVAILABLE FOR BOX CONSTRUCTION

STANDARD DEFECTS AND BLEMISHES IN LUMBER

Commercial lumber grades are based on the number, size and position of the defects and blemishes the lumber contains, and, to a certain extent, on the size of the pieces.

Defect—A defect is any irregularity occurring in or on wood that may lower some of its strength values.

Blemish—A blemish is anything not classified as a defeet which mars the appearance of the wood.

The following are the principal recognized defects and blemishes in lumber which may lower its grade[1]. (See also Plate I.)

1. **Knots**—Knots are irregularities of growth caused by the junction of a branch with the body of a tree. These are further classified as to size, shape, and quality.

A *pin knot* is less than ½ inch in diameter.

A *standard knot* is from ½ to 1½ inches in diameter.

[1]The definitions of defects as here given are not universally accepted. For example, a standard knot, according to the West Coast Lumbermen's Association, is from ¾ to 1½ inches in diameter; under the rules of the Southern Cypress Manufacturers' Association a standard knot is from ¾ to 1¼ inches in diameter; and according to the two hardwood associations a knot 1¼ inches in diameter is considered a standard defect. At the present time, standard definitions of terms relating to defects and blemishes are being revised by the National Association of Lumber Manufacturers.

A *large knot* is over 1½ inches in diameter

A *round knot* is circular or oval in form.

A *spike knot* is cut through lengthwise, and, therefore, can occur on quarter-sawed surfaces only.

A *sound knot* is as hard as the wood it is in, and is so fixed by growth or position that it will retain its place in the piece.

An *encased knot* is not firmly connected throughout with the surrounding wood. If intergrown partially with the surrounding wood or so held by shape or position that it will retain its place in the piece, it is considered a sound knot.

A *water-tight knot* is completely intergrown with the surrounding wood on one face, and is sound on that face.

A *loose knot* is not firmly held in position; it may drop out.

A *pith knot* is a sound knot with a hole not over ¼ inch in diameter at the center.

A *rotten knot* is not so hard as the wood it is in.

2. **Shake**—Shake is a partial or entire separation of the wood between annual rings.

3. **Checks**—Checks are splits which run radially across the rings. They are usually due to unequal shrinkage in seasoning.

4. **Splits**—Splits are due to rough handling or internal stresses and are easily confused with checks and shakes.

5. **Pitch Pockets**—Pitch pockets are lens-shaped openings between the annual rings of some conifers and contain more or less pitch or bark.

6. **Pitch Streaks**—Pitch streaks are conspicuous accumulations of pitch in the wood cells.

7. **Wane**—Wane is bark on the edge of a piece or the absence of the square edge.

8. **Rot and Dote**—Rot and dote refer to different stages of decay due to wood-destroying fungi. Either is permissible to a certain extent in the lower grades of lumber.

9. **Stain**—Stain (as blue stain, brown stain, water stain, and others not due to decay) usually affects only the appearance of lumber and is not considered a defect in the lower grades.

10. **Pith**—Pith is the small soft core at the structural center of a log. It is often surrounded by small checks,

shakes, numerous pin knots, etc. In some woods it is large enough to be objectionable on the face of lumber.

11. **Worm Holes**—Worm holes are very common in some woods and may render them unfit for high-grade work. They may be small, in which case they are known as pin-worm holes, or if in groups, shot-worm holes; or they may be large, in which case they are known as grub-worm holes.

12. **Bird Pecks and Gum Spots**—Bird pecks and gum spots appear as brown or black discolorations. Sometimes an open cavity is formed, which in the case of gum spots may contain a gum-like substance.

13. **Rafting Pin Holes**—Rafting pin holes may be bored for rafting pins or holes made by driving rafting pins into the wood.

14. **Warping**—Warping includes both twisting and cupping and is due mostly to improper piling and drying methods and may cause clear lumber to be put in a low grade.

15. **Crook**—A crook is the curving of a piece edgewise in the longitudinal direction.

16. **Bow**—A bow is the deviation of a piece flatwise

17. **Twisting**—Twisting is the turning or winding of the edges of a piece so that the four corners of a face are no longer in the same plane.

18. **Cupping**—Cupping is the curving of a piece across the grain or width of the piece.

19. **Poor Manufacture**—Poor manufacture, as irregular width or thickness or chipped or torn grain, is cause for putting lumber in a lower grade.

GRADING ACCORDING TO SIZE OF LUMBER

Standard Sizes of Lumber—The sizes of lumber available to the box manufacturer are, in general, the same as are made for the general lumber trade. The thicknesses used are normally in the rough, 4/4, 5/4, 6/4, 7/4, and 8/4 inches. These thicknesses are resawed in the box factory to produce 1/4, 5/16, 3/8, 7/16, 1/2, 9/16, 5/8, 11/16, 3/4, and 13/16-inch material surfaced on one side or scantily surfaced on two sides. The widths range from about 3 or 4 inches up to about 12 inches, or sometimes up to 16 inches. In some woods, stock widths only are manufactured, that is, even-inch widths (4, 6, 10, and 12, etc., inches); in other woods, both stock and random sizes (odd inch and fractional inch widths) are made; in hardwoods, random widths only are standard, but they are measured to the nearest whole inch. Lengths of box lumber generally range from 1 to 20 feet. Regular grades are often cut from 6 or 8 feet up to 16 or 20 feet; while a special grade, usually known as short box, includes the shorter lengths.

Much of the lumber cut in the New England States is not edged at the mill but is put on the market in the form of tapering boards with waney edges. This is known as "round-edged" lumber and consists mostly of second-growth white pine, spruce, hemlock, and fir. About 10 per cent more box lumber can be cut from New England timber if it is not edged until after it is cut to lengths and if it is not edged to standard widths. On account of the difficulty of satisfactorily grading round-edged lumber, it is not graded, as a rule, but is sold as long run.

Grades Suitable for Boxes and Crates—Grading rules for a certain species or group of woods are prepared by the lumbermen's associations particularly interested in that kind of lumber. At the present time there are a number of lumber associations which have standard grading rules. Some woods are graded by more than one association under rules which are not similar, a condition which causes more or less confusion in preparing lumber. Neither are the names, qualities, or sizes of similar grades of the different associations always alike.

The upper grades, which contain fewest defects and a very limited per cent of narrow widths and short lengths, are seldom used in the manufacture of shipping containers. The low grades, which contain more or less knots and other defects, furnish the material commonly used.

The problem of the manufacturer is to cut out the defects not permissible in the kind of boxes he is making and to do this with as little waste as possible and with the least expenditure of power and labor. In general, the waste of lumber in making boxes is from 15 to 20 per cent. Insistence upon boxes finished without knots would result in a much larger waste or in the use of a finishing grade. Because of the cost of lumber and labor, box specifications should permit the use of low grades and cause as little waste in working up the lumber as is consistent with the box requirements.

TABLE 4. APPROXIMATE WEIGHTS OF LUMBER PER CUBIC FOOT, AND PER SQUARE FOOT OF USUAL THICKNESSES USED IN PACKING BOXES, THOROUGHLY AIR-DRY (12 TO 15 PER CENT), AND ESTIMATED SHIPPING WEIGHTS

Species	Specific gravity oven-dry weight green volume	Pounds per cu. ft. air-dry [1]	Actual thickness in inches (Pounds per square ft.)													Pounds per M board feet of rough 1 inch or more in thickness "shipping dry"
			1	7/8	13/16	3/4	5/8	1/2	3/8	5/16	1/4	3/16	1/6	1/8	1/10	
			1.83	1.60	1.49	1.37	1.14	0.91	0.69	0.57	0.46	0.34	0.30	0.23	0.19	
Very light:																
Arborvitae	.293	22	1.83	1.60	1.49	1.37	1.14	.91	.69	.57	.46	.34	.30	.23	.19	below 2300
Light:																
Alpine fir	.306	23	1.92	1.68	1.56	1.44	1.20	.96	.72	.6	.48	.36	.32	.24	.19	
Western red cedar	.310	22	1.83	1.60	1.49	1.37	1.14	.91	.69	.57	.46	.34	.30	.23	.18	
Engelmann spruce	.312	24	2.00	1.75	1.63	1.50	1.25	1.00	.75	.62	.50	.38	.33	.25	.20	
Red fir	.322	28	2.33	2.04	1.89	1.75	1.46	1.16	.87	.73	.58	.44	.39	.29	.23	
Basswood	.325	26	2.17	1.90	1.76	1.63	1.36	1.08	.81	.68	.54	.41	.36	.27	.22	
Yellow buckeye	.326	25	2.08	1.82	1.69	1.56	1.30	1.04	.78	.65	.52	.39	.35	.26	.21	2300 to 2600
White fir	.332	24	2.00	1.75	1.63	1.50	1.25	1.00	.75	.62	.50	.38	.33	.25	.20	
Balsam fir	.335	25	2.08	1.82	1.69	1.56	1.30	1.04	.78	.65	.52	.39	.35	.26	.20	
Sitka spruce	.341	26	2.17	1.90	1.76	1.63	1.36	1.08	.81	.65	.52	.39	.35	.27	.21	
Black willow	.348	26	2.17	1.90	1.76	1.63	1.36	1.08	.81	.68	.54	.41	.36	.27	.22	
White fir	.350	26	2.17	1.90	1.76	1.63	1.36	1.08	.81	.68	.54	.41	.36	.27	.22	
Noble fir	.350	26	2.17	1.90	1.76	1.63	1.36	1.08	.81	.68	.54	.41	.36	.27	.22	
Butternut	.359	26	2.17	1.90	1.76	1.63	1.36	1.08	.81	.68	.54	.41	.36	.27	.22	
Moderately light:																
Sugar pine	.360	26	2.17	1.90	1.76	1.63	1.36	1.08	.81	.68	.54	.41	.36	.27	.22	
Aspen	.360	28	2.33	2.04	1.89	1.75	1.46	1.16	.87	.73	.58	.44	.39	.29	.23	
White spruce	.360	28	2.33	2.04	1.89	1.75	1.46	1.16	.87	.73	.58	.44	.39	.29	.23	
White spruce	.363	27	2.25	1.97	1.83	1.69	1.41	1.12	.84	.70	.56	.42	.38	.28	.22	
Red fir	.370	27	2.25	1.97	1.83	1.69	1.41	1.12	.84	.70	.56	.42	.38	.28	.22	
Yellow poplar	.371	27	2.25	1.97	1.83	1.69	1.41	1.12	.84	.70	.56	.42	.38	.28	.22	
Red (non)	.372	28	2.33	2.04	1.89	1.75	1.46	1.16	.87	.73	.58	.44	.39	.29	.23	
Amabilis fir	.373	27	2.25	1.97	1.83	1.69	1.41	1.12	.84	.70	.56	.42	.38	.28	.22	
Red spruce	.380	28	2.33	2.04	1.89	1.75	1.46	1.16	.87	.73	.58	.44	.39	.28	.22	2600 to 3000
Lodgepole p. in.	.380	28	2.33	2.04	1.89	1.75	1.46	1.16	.87	.73	.58	.44	.39	.29	.23	
Western yellow pine	.382	28	2.33	2.04	1.89	1.75	1.46	1.16	.87	.73	.58	.44	.39	.29	.23	
Eastern hemlock	.383	28	2.33	2.04	1.89	1.75	1.46	1.16	.87	.73	.58	.44	.39	.29	.23	
Western white pine	.393	29	2.42	2.12	1.97	1.82	1.51	1.21	.91	.76	.60	.45	.40	.30	.24	
Jack pine	.394	29	2.42	2.12	1.97	1.82	1.51	1.21	.91	.76	.60	.45	.40	.30	.24	
Noble fir	.396	30	2.50	2.19	2.03	1.88	1.56	1.25	.94	.78	.62	.47	.42	.31	.25	
Western hemlock	.398	28	2.33	2.04	1.89	1.75	1.46	1.16	.87	.73	.58	.44	.39	.29	.23	
Douglas fir (mountain type)	.401	30	2.50	2.19	2.03	1.88	1.56	1.25	.94	.78	.62	.47	.42	.31	.23	
Redwood	.410	30	2.50	2.19	2.03	1.88	1.56	1.25	.94	.78	.62	.47	.42	.31	.25	
Port Orford cedar	.411	31	2.58	2.26	2.10	1.94	1.61	1.29	.97	.81	.64	.48	.43	.32	.26	
Bald cypress	.412	30	2.50	2.19	2.03	1.88	1.56	1.25	.94	.78	.62	.47	.42	.31	.25	

[1]

TABLE 4. APPROXIMATE WEIGHTS OF LUMBER PER CUBIC FOOT, AND PER SQUARE FOOT OF USUAL THICKNESSES USED IN PACKING BOXES, THOROUGHLY AIR-DRY (12 TO 15 PER CENT MOISTURE), AND ESTIMATED SHIPPING WEIGHTS—Concluded

Species	Specific gravity oven-dry weight green volume	Pounds per cu. ft. air-dry[1]	Actual thickness in inches (Pounds per square foot)												Pounds per M board feet of rough lumber 1 inch or more in thickness "shipping dry"
			1	7/8	13/16	3/4	5/8	1/2	3/8	5/16	1/4	3/16	1/8	1/10	
Moderately heavy:															
White elm	.437	35	2.92	2.56	2.37	2.19	1.82	1.46	1.10	.91	.73	.55	.36	.29	3000 to 3500
River maple	.439	32	2.67	2.34	2.17	2.00	1.67	1.34	1.00	.83	.67	.50	.33	.27	
Scrub oak		33²	2.75	2.41	2.23	2.06	1.72	1.38	1.03	.86	.69	.52	.34	.28	
Cucumber tree	.440	33	2.75	2.41	2.23	2.06	1.72	1.38	1.03	.86	.69	.52	.34	.28	
Gray gum	.440	34	2.83	2.48	2.30	2.12	1.77	1.42	1.06	.88	.71	.53	.35	.28	
Red gum	.443	34	2.83	2.48	2.30	2.12	1.77	1.42	1.06	.88	.71	.53	.35	.28	
Douglas fir (Pacific coast type)	.444	34	2.83	2.48	2.30	2.12	1.77	1.42	1.06	.88	.71	.53	.35	.28	
Oregon gum oak	.455	34	2.83	2.48	2.30	2.12	1.77	1.42	1.06	.88	.71	.53	.35	.28	
Sycamore	.456	34	2.83	2.48	2.30	2.12	1.77	1.42	1.06	.88	.71	.53	.35	.28	
Black ash	.458	34	2.83	2.48	2.30	2.12	1.77	1.42	1.06	.91	.71	.53	.35	.28	
Magnolia (evergreen)	.460	35	2.92	2.56	2.37	2.19	1.82	1.46	1.10	.91	.73	.55	.36	.29	
Black gum	.462	35	2.92	2.56	2.37	2.19	1.82	1.46	1.10	.91	.73	.55	.36	.29	
Fish pine	.470	35	2.92	2.56	2.37	2.19	1.82	1.46	1.10	.91	.73	.55	.36	.29	
Paper birch	.473	38	3.17	2.77	2.58	2.38	1.98	1.58	1.19	.99	.79	.59	.40	.32	
Western larch	.481	36	3.00	2.63	2.44	2.25	1.88	1.50	1.12	.94	.75	.56	.38	.30	
Pumpkin ash	.485	36	3.00	2.63	2.44	2.25	1.88	1.50	1.12	.94	.75	.56	.38	.30	
Red oak	.485	37	3.08	2.70	2.50	2.31	1.92	1.54	1.15	.96	.77	.58	.38	.31	
Slippery elm	.485	37	3.08	2.70	2.50	2.31	1.92	1.54	1.15	.96	.77	.58	.38	.31	
Hackberry	.486	37	3.08	2.70	2.50	2.31	1.92	1.54	1.15	.96	.77	.58	.38	.31	
Tamarack	.491	37	3.08	2.70	2.50	2.31	1.92	1.54	1.15	.96	.77	.58	.38	.31	
Shortleaf pine	.494	38	3.17	2.77	2.58	2.38	1.98	1.58	1.19	.99	.79	.59	.40	.32	
Heavy:															
Loblolly pine	.504	38	3.17	2.77	2.58	2.38	1.98	1.58	1.19	.99	.79	.59	.40	.32	3500 to 4000
White ash	.523	40	3.33	2.91	2.71	2.50	2.08	1.66	1.25	1.04	.83	.62	.42	.33	
Beech	.544	44	3.67	3.21	2.98	2.75	2.29	1.84	1.38	1.15	.92	.69	.46	.37	
Yellow birch	.550	44	3.67	3.21	2.98	2.75	2.29	1.84	1.38	1.15	.92	.69	.46	.37	
Longleaf pine	.551	42	3.50	3.06	2.84	2.63	2.19	1.75	1.31	1.09	.88	.66	.44	.35	
Sugar maple	.560	43	3.58	3.13	2.91	2.68	2.24	1.79	1.35	1.12	.89	.67	.45	.36	
Red oak	.563	44	3.67	3.21	2.98	2.75	2.29	1.84	1.38	1.15	.92	.69	.45	.37	
Cork (rock) elm	.574	44	3.67	3.21	2.98	2.75	2.29	1.84	1.38	1.15	.92	.69	.46	.37	
Sweet birch	.588	45	3.75	3.28	3.05	2.81	2.34	1.88	1.41	1.17	.94	.70	.47	.38	
Commercial white oak	.589	47	3.92	3.43	3.18	2.94	2.45	1.96	1.47	1.22	.98	.73	.49	.39	

[1] The apparent discrepancies when specific gravity and weight per cubic foot in some cases are due to the fact that the specific gravity is based on the green volume and the weight per cubic foot on the air-dry volume.
² from tenth column, all data from Forest Service data.

IMPORTANT PHYSICAL PROPERTIES OF WOOD WHICH INFLUENCE ITS USE IN BOX CONSTRUCTION

WEIGHT

Importance—The weight of the wood used for packing boxes and crates is very important. It influences the cost of both handling and transportation. The strength, shrinking, and warping of lumber, and the ease with which it splits in nailing increase, as a rule, with the dry weight. Where strength is an important factor, light pieces, no matter what the species may be, should not be used. Thinner pieces may be used of the heavier woods than of the lighter woods, without reducing the strength. The denser woods hold nails better and are desirable on this account. On the other hand, the lighter woods give less trouble in seasoning and manufacture.

How the Weight of Lumber is Expressed—In commercial practice the weight of lumber is usually expressed in pounds per thousand board feet when "shipping-dry." This ranges from about 2,100 pounds for very light woods to over 4,000 pounds for very heavy woods. A more definite way of expressing the weight of wood is in pounds per cubic foot or per square foot of specified thickness at any given moisture content or degree of seasoning, as green, thoroughly air dry, kiln dry, or oven dry. For the convenience of box designers and manufacturers the approximate weights for different thicknesses of box lumber and veneer are given in Table 4.

The weight is often expressed in terms of specific gravity, which means the ratio of the weight of an object to the weight of an equal volume of water.[1] To determine the specific gravity of wood, it is customary to use the oven-dry weight and the volume measured while the wood is in the same condition or when green or partly dry. It should always be stated under what moisture condition the weight and volume were measured. Table 4 gives the specific gravity of box woods.

FACTORS WHICH INFLUENCE THE WEIGHT OF LUMBER

1. **Species**—A great variation is found in the weight of the various commercial species, as can be seen from Table 4.

---

[1] A cubic foot of water weighs approximately 62.5 pounds.

2. Density, or Amount of Wood Substance—Even in the same species there is considerable variation in weight due to differences in density (the figures given in Table 4 are only average, or approximate). Species growing in the wet swamps of the South show the greatest variation. The swelled butt logs of trees growing in places where the ground is covered with water a large part of the year usually contain very light wood. Higher up in the tree the wood is denser and heavier. Very light pieces of cotton gum (tupelo), cypress, and ash usually come from such localities. Ordinarily, however, butt logs produce the heaviest wood.

3. Moisture Content—The moisture in kiln-dry wood adds only about 5 or 10 per cent (occasionally more) to the weight, but in green wood the water contained may weigh more than the wood itself. Thoroughly air-dry wood usually contains from 12 to 15 per cent moisture.

4. Resin Content—In yellow pine, Douglas fir, tamarack, and occasionally in spruce, parts of the tree trunk become infiltrated with resin to a considerable extent. "Fatty pieces," as such resinous pieces are called, are considerably heavier than normal wood.

Moisture Content

Importance—A knowledge of how much moisture is contained in wood when manufactured and when put into use is exceedingly important. When boxes or shooks are manufactured under certain moisture conditions and then stored in a warehouse or shipped to a drier or wetter climate, the moisture content will accommodate itself to the varied atmospheric conditions. This affects the shrinking or swelling, warping, checking, weight, strength, and nail-holding power of the wood.

How Determined—To determine the approximate moisture content of a stack of box material:

Select one representative piece from every 100 or 500 pieces. In the case of lumber, sections ½ inch or less with the grain should be cut two or more feet from the end at a place free from knots, rot, or other abnormalities, and where sapwood and heartwood are in representative proportions.

Immediately after cutting these sections, pick off all loose slivers and weigh the samples to an accuracy of one-half of

1 per cent. If a delicate scale is not available, several sections may be taken out of each piece to insure greater accuracy. This weight we will call the *original weight.*

Dry the sections in an oven in which an even temperature of about 212° F. and free circulation of air over end grain can be maintained, until they no longer lose in weight. Sections ½ inch long will usually dry out completely over night. If a drying oven is not available, the samples will reach within 1 or 2 per cent of the same dryness when laid on pipes containing live steam.

Weigh the dry samples to an accuracy of one-half of 1 per cent and subtract this *oven-dry weight* from the original weight. Divide the difference by the oven-dry weight and multiply by 100. This gives the per cent moisture *based on the oven-dry weight.*

Thus, if the original weight is 625.7 grams and the oven-dry weight 438.2 grams, the moisture content is ·

$$625.7 - 438.2 = 187.5 \text{ grams; and}$$

$$\frac{187.5}{438.2} \times 100 = 42.8 \text{ per cent moisture content}$$

Variation—In green timber the moisture content ranges from about 30 to 250 per cent, based on oven-dry weight. In so called air-dry wood it may range from 5 or 8 per cent, as in small pieces, to over 30 per cent, as in timber dried to reduce its shipping weight. Kiln-dry wood is also highly variable in moisture content, depending on the purpose for which it is dried and the care with which the drying is carried on. If the object of the drying is to reduce quickly the shipping weight, the lumber may be no drier or not even so dry as it would become from prolonged air drying. On the other hand, lumber may become too dry in a kiln and give trouble during and after manufacture.

How the Moisture is Contained in Wood—The troubles from shrinking, swelling, warping, and cupping of lumber when put into use arise from the manner in which the moisture is held in the wood. It is held principally in two ways, (1) filling the cell cavities and (2) absorbed in the cell walls. When wood dries, the moisture first leaves the cell cavities, and after they are empty it begins to leave the cell walls. The condition in which the cell cavities are empty but the cell walls still fully saturated is known as the "fiber-saturation point." As a rule wood does not shrink in drying or lose in strength until the moisture content falls below the fiber-

saturation point. This moisture content ranges from 20 to 35
per cent. The moisture content of seasoned lumber does not
remain constant but varies with the humidity of the surrounding atmosphere. This is due to the cell walls giving off or
absorbing moisture. · (The curve in figure 2 shows the mois

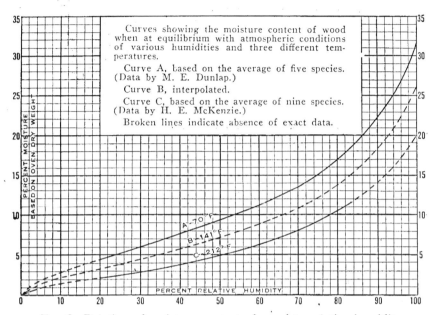

Curves showing the moisture content of wood
when at equilibrium with atmospheric conditions
of various humidities and three different temperatures.

Curve A, based on the average of five species.
(Data by M. E. Dunlap.)

Curve B, interpolated.

Curve C, based on the average of nine species.
(Data by H. E. McKenzie.)

Broken lines indicate absence of exact data.

Fig. 2—Relation of moisture content of wood to relative humidity.

ture content at which wood will ultimately arrive under a
given humidity and various temperatures. It is based on only
a limited number of woods but is believed to be representative for most species.) This relation of the moisture content
of wood to the relative humidity is of great significance in
the manufacture and use of wooden articles.

Proper Moisture Content of Box Lumber—The moisture
content of wood for any purpose should be, at the time of
manufacture, approximately what it will be when the wood is
in use. For boxes and crates from 12 to 18 per cent is considered a safe approximation. ·

If the lumber used has too high a moisture content various weaknesses develop later (see page 47); the box will be
heavier than necessary; if it is made for a standard-sized ar-
ticle, it will not fit the contents unless allowance is made for
shrinking; and if it is not assembled immediately, the different parts may not fit together properly.

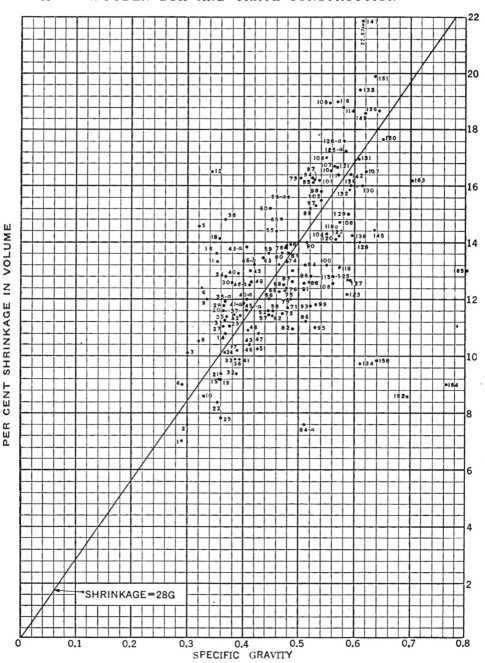

Fig. 3—Relation between the volumetric shrinkage and specific gravity of various American woods. See opposite page for list of species and reference numbers.

HARDWOODS

Species	Locality	Ref. No.
Alder, red	Wash.,	30
Ash, biltmore	Tenn.	91
black	Mich.	60
black	Wis.	70
blue	Ky.	99
green	La.	93
green	Mo.	100
pumpkin	Mo.	79
white	Ark.	106
white	N. Y.	128
white	W. Va.	83
Aspen	Wis.	23
largetooth	Wis.	20
Basswood	Pa.	12
"	Wis.	5
Beech	Ind.	110
"	Pa.	98
Birch, paper	Wis.	73
sweet	Pa.	129
yellow	Pa.	107
yellow	Wis.	103
Buckeye, yellow	Tenn.	9
Buckthorn, cascara	Ore.	84a
Butternut	Tenn.	27
"	Wis.	21
Chinquapin, Western	Ore.	46b
Cherry, black	Pa.	72
Cherry, wild red	Tenn.	24
Chestnut	Md.	46
"	Tenn.	40
Cottonwood, black	Wash.	6
Cucumber tree	Tenn.	59
Dogwood (flowering)	Tenn.	151
(Western)	Ore.	125a
Elder, pale	Ore.	69a
Elm, cork	Wis., Marathon Co.	126
"	Wis., Rusk Co.	120
Elm, slippery	Ind.	102
slippery	Wis.	74
white	Pa.	55
white	Wis.	53
Greenheart		165
Gum, black	Tenn.	68
blue (eucalyptus)	Cal.	147
cotton	La.	76
red	Mo.	54
Hackberry	Ind.	90
"	Wis.	78
Haw, pear	Wis.	146
Hickory, big shellback	Miss.	135
big shellback	Ohio	154
Hickory, bitternut	Ohio	139
mockernut	Miss.	144
mockernut	Pa.	159
mockernut	W. Va.	155
nutmeg	Miss.	112
pignut	Miss.	148
pignut	Ohio	157
pignut	Pa.	160
pignut	W. Va.	161
shagbark	Miss.	140
shagbark	Ohio	152
shagbark	Pa.	143
shagbark	W. Va.	153
water	Miss.	141
Holly, American	Tenn.	87
Hornbeam	Tenn.	149
Laurel, mountain	Tenn.	145
Locust, black	Tenn.	158
honey	Ind.	162
Madrona	Cal.	101
"	Ore.	128a
Magnolia	La.	66
Maple, Oregon	Wash.	58
red	Pa.	69
red	Wis.	92
silver	Wis.	56
sugar	Ind.	104
sugar	Pa.	108
sugar	Wis.	124
Oak, burr	Wis.	125
California black	Cal.	80
canyon live	Cal.	163
chestnut	Tenn.	121
cow	La.	133
laurel	La.	116
post	Ark.	130
post	La.	137
red	Ark.	119

Species	Locality	Ref. No.
Oak. red	Ind.	118
red	La.	117
red	Tenn.	97
Highland Spanish	La.	94
Lowland Spanish	La.	142
swamp white	Ind.	150
tanbark	Cal.	115
water	La.	111
white	Ark.	132
white	Ind.	138
white	La., Richland Parish	136
white	La., Winn Parish	131
willow	La.	109
yellow	Ark.	122
yellow	Wis.	105
Osage orange	Ind.	164
Poplar, yellow (tulip-tree)	Tenn.	35
Rhododenron, great	Tenn.	85
Sassafras	Tenn.	51
Serviceberry	Tenn.	156
Silverbell-tree	Tenn.	49
Sourwood	Tenn.	89
Sumac, staghorn	Wis.	61
Sycamore	Ind.	63
"	Tenn.	65
Umbrella, Fraser	Tenn.	45
Willow, black	Wis.	11
Willow, Western black	Ore.	43a
Witch hazel	Tenn.	114

CONIFERS

Species	Locality	Ref. No.
Cedar, incense	Cal.	26
Western red	Mont.	2
Western red	Wash.	10
white	Wis.	1
Cypress. bald	La.	62
Fir, Alpine	Colo.	4
amabilis	Ore.	39
amabilis	Wash.	18
balsam	Wis.	14
Douglas	Cal.	45a
Douglas	Ore.	67a
Douglas	Wash. and Ore.	67
Douglas	Wash., Lewis Co.	75
Douglas	Wash., Chehalis Co.	46a
Douglas	Wyo.	48
grand	Mont.	36
noble	Ore.	16
white	Cal.	17
Hemlock, black	Mont.	47
Eastern	Tenn.	52
Eastern	Wis.	15
Western	Wash.	50
Larch, Western	Mont.	84
Western	Wash.	64
Pine, Cuban	Fla.	127
jack	Wis.	43
Jeffrey	Cal.	33
loblolly	Fla.	88
lodgepole	Colo.	31
lodgepole	Mont., Gallatin Co.	35a
lodgepole	Mont., Granite Co.	41a
lodgepole	Mont., Jefferson Co.	40a
lodgepole	Wyo.	34
longleaf	Fla.	123
longleaf	La., Tangipahoa Parish	96
longleaf	La., Lake Charles	113
longleaf	Miss.	95
Norway	Wis.	57
pitch	Tenn.	71
pond	Fla.	86
shortleaf	Ark.	77
sugar	Cal.	22
Table Mountain	Tenn.	82
Western white	Mont.	42
Western yellow	Ariz.	19
Western	Cal.	37
Western	Colo.	41
Western	Mont.	32
white	Wis.	25
Redwood	Cal., Albion.	28
"	Cal., Korbel.	13
Spruce, Engelmann	Colo., San Micuel Co.	3
Engelmann	Colo., Grand Co.	8
red	N. H.	44
red	Tenn.	29
white	N. H.	7
white	Wis.	38
Tamarack	Wis.	81
Yew, Western	Wash.	134

If the lumber is too dry, it splits more easily during manufacture and in nailing, and is slightly more difficult to work; and, although the strength of the wood is greater, the assembled box is considerably weaker than one of the proper moisture content.[1]   Furthermore, the wood will swell later,

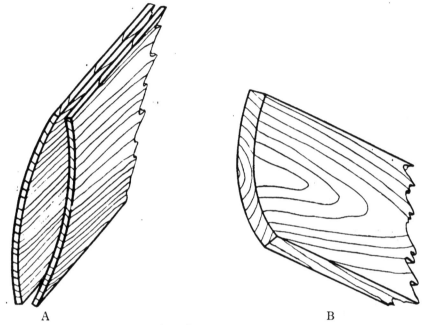

Fig. 4—Cupping of lumber

A. Cupping of the two halves of a resawed board.
B. Cupping of plain-sawed lumber while seasoning.

bulging and warping, which will result in extra strain on the nails, sometimes bending or withdrawing them. If the shooks are stored in a knock-down condition they may not fit closely together when later assembled.

Dry wood is stronger in most respects than green wood of the same quality. The increase begins as soon as the wood dries to below the fiber-saturation point. No allowance, however, for increase in strength above that of green material should be made for large dimension stock used for skids for crates, etc., because very often these are not below the fiber-saturation point in the interior and seasoning checks may develop which counteract any increased strength of the fibers due to seasoning.

---

[1]This is discussed in greater detail in Chapter II, page 47.

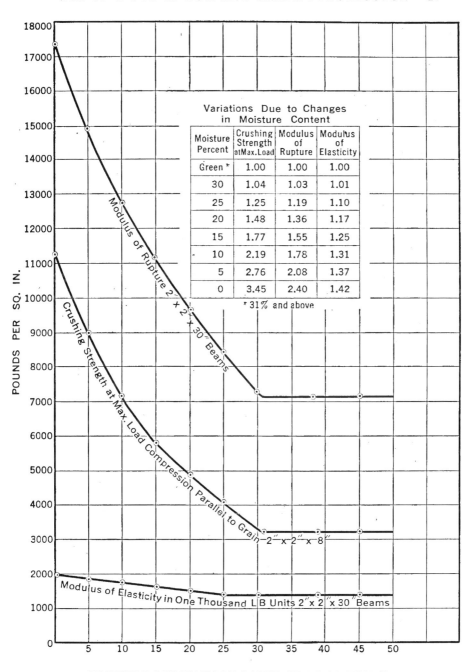

Variations Due to Changes in Moisture Content			
Moisture Percent	Crushing Strength atMax.Load	Modulus of Rupture	Modulus of Elasticity
Green *	1.00	1.00	1.00
30	1.04	1.03	1.01
25	1.25	1.19	1.10
20	1.48	1.36	1.17
15	1.77	1.55	1.25
10	2.19	1.78	1.31
5	2.76	2.08	1.37
0	3.45	2.40	1.42

* 31% and above

MOISTURE PERCENTAGE BASED ON DRY WEIGHT

FIG. 5—Effect of moisture upon the strength of small clear specimens of western hemlock.

The various strength properties of wood do not increase in the same proportion as wood seasons. This is shown for western hemlock in figure 5. In the case of shear along the grain, wood very often fails to show large increase in strength as it dries, probably because of the checks which form in shrinking. The shock resisting ability shows no appreciable increase in most woods and decreases slightly in many species while seasoning.

There is more or less prejudice against kiln-dried lumber for use where strength is essential. There is no doubt that a large amount of lumber is damaged by improper methods of kiln drying. However, properly kiln-dried material is just as strong as similar lumber air-dried to the same moisture content, and may be even stronger.

Shrinking and Swelling of Wood

Extent—When green or soaked wood dries it does not shrink until it gets down to about 25 to 30 per cent moisture (fiber-saturation point) and from there on it shrinks until the oven-dry condition is reached. Conversely, when dry wood absorbs moisture it swells until the fiber-saturation point is reached and beyond that there is no more change in dimensions, although absorption may continue, increasing the weight. Therefore, about half of the total possible shrinkage has taken place when wood is seasoned down to 12 or 15 per cent moisture, which corresponds to the thoroughly air-dry condition (see figure 6). Lumber which is only partly seasoned or which is "shipping-dry" when received in the factory, may not have shrunk appreciably, and considerable shrinkage may take place during and after manufacture.

The total shrinkage from the green to the oven-dry condition varies greatly for the different species of woods, but in general it increases with the weight of the wood. Figure 3 shows the relation of shrinkage in volume to specific gravity.

The shrinkage along the grain is so slight as to be negligible for most purposes. Across the grain, however, it is considerable; and it is decidedly less radial than tangential in the same piece of wood. This is shown in Table 5.

Occasionally, what appears to be abnormal shrinkage takes place in drying certain kinds of lumber. The surfaces of such lumber have a caved-in or corrugated appearance when dried. (See figure 7.) Shrinkage in some species is more likely to occur when the wood is kiln-dried from the saw at high temperature. In such cases, some of the cells of the wood collapse as the water leaves them. This does not

TABLE 5. PER CENT OF SHRINKAGE[1] ACROSS THE GRAIN

	From green or over 30 per cent moisture to the oven-dry condition.		From green or over 30 per cent moisture to the air-dry condition — 12 to 15 per cent.	
	Very light woods	Very heavy woods	Very light woods	Very heavy woods
Radial, i. e., across the rings	2.6	6.3	1.3	3.1
Tangential, i. e., along the rings	4.7	11.2	2.3	5.6

occur in the sapwood or at the ends or edges of lumber where air can readily enter the wood and prevent collapse of the cells.

FIG. 6—End of honeycombed oak plank.

HOW TROUBLES FROM SHRINKING AND SWELLING MAY BE REDUCED TO A MINIMUM IN BOXES

There is practically no method of treatment by means of which the shrinking and swelling of wood, when exposed to varying atmospheric conditions, can be entirely overcome; but if the following precautions are observed so far as possible in the selection and treatment of wood, trouble from these sources will be reduced to a minimum.

1. Select light woods if other requirements permit.

2. Be sure the wood is dried to the proper moisture content.

3. Use quarter-sawed lumber—this is practical only for special boxes.

---

[1]Shrinkage is here expressed in per cent of green dimension.

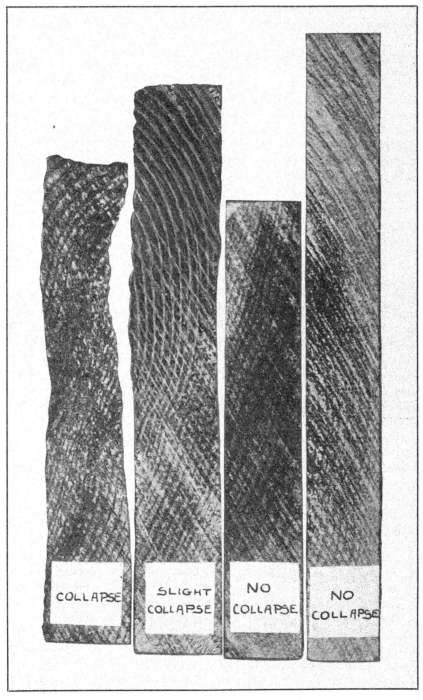

FIG. 7—Collapse in 1-inch boards.

4. Use plywood, that is, thin sheets of veneer glued together with the grain crossed.

5. Cover the wood with oil, paint, or other protective coating.

6. Avoid as far as possible storage or use of boxes under widely varying atmospheric conditions

Checking—Checking in wood is due to stresses set up on account of uneven shrinkage. End checking is very common and often can not be avoided. It is caused by the wood's drying more rapidly at the ends than some distance from the ends, where drying takes place only from the sides.

Checks on the face of lumber, known as surface checks, are due to the surface drying much more rapidly than the interior. Wood which is badly surface-checked splits more easily in machining and nailing. Timbers containing the pith will invariably check in seasoning because the shrinkage in the circumferential direction is greater than toward the center.

Cupping—By cupping is meant the curvature of lumber across the grain, which gives it more or less of a trough-like appearance. It may be due to one side drying more rapidly than the other, in which case it is temporary. Permanent cupping takes place in plain-sawed lumber when dried with insufficient weight on it. In plain-sawed lumber the side toward the center of the tree shrinks less in width, thus causing the lumber to curve away from the center as it dries, as illustrated in figure 4.

Casehardening—By casehardening is meant a condition of internal stress in seasoned lumber which causes it to cup inwardly when resawed. (See figure 4B.) Since much box lumber is resawed, casehardening gives considerable trouble in the box industry.

Honeycombing—In some casehardened lumber the internal stress becomes so great that the wood is torn apart, producing internal checks known as "honeycomb." (See figure 6.) These checks also extend along the medullary rays and may come to the surface.

Color—The manufacturer of boxes and crates can not pay much attention to the color of the wood he uses. Light colored woods are usually preferable, however, because addresses and advertising matter show up better than on dark woods. This is one reason why pine leads as a box material. For certain high-grade containers only white woods are used.

Odor and Taste—Containers of certain kinds of food must be free from odor or taste or they will taint the contents. It is not the purpose of this publication to discuss which foods are and which are not easily tainted. The following species of wood have a pronounced odor and should not be used for shipping certain classes of food: all of the cedars (including arborvitae), Alpine fir, yellow pine, and sassafras

MECHANICAL OR STRENGTH PROPERTIES OF WOOD[1]

Meaning of Strength—Strength in the broad sense of the word is the summation of the mechanical properties of a material, or its ability to resist stress or deformation. While such properties as hardness, stiffness, and toughness are not always thought of in connection with the term "strength,"[2] they are unconsciously included when in a specific instance they are important. Such expressions as strength in shear, strength in compression, and strength as a column are very specific and allow little chance for confusion. (See Tables 6 and 7.)

Tensile Strength—The tensile strength of a material is measured by the resistance it offers to forces which tend to pull it apart. In wood, tension may be produced along the grain or across the grain. The tensile strength along the grain is many times greater than it is across the grain. It is almost impossible in ordinary construction to develop full strength in tension along the grain since the fastenings are usually inadequate; for this reason tension tests along the grain are seldom made.

Compression Strength—The strength in compression is measured by the resistance a material offers to forces which tend to crush it. In wood, these forces may act along the grain or at right angles to it. In compression parallel to the grain, as in a post, wood shows great strength. In compression perpendicular to the grain, no maximum load is reached; crushing takes place as the load is increased. Crushing strength is important in determining the size of bearing areas in heavy crates.

Shearing Strength—By shearing strength is meant the resistance a piece of wood offers to a force which tends to

---

[1]Various values have been combined and are given in Table 7, as strength as a beam or post, stiffness, shock resisting ability, and hardness. The basis for deriving these values is given in Table 6.

[2]Methods of determining the strength values of wood are explained in Bulletin 556, Forest Service, U. S. Department of Agriculture.

slide one portion of it over the other portion. It varies as the area of the plane along which the shear occurs. Boxes and

TABLE 6. PHYSICAL AND MECHANICAL PROPERTIES OF WOODS GROWN IN THE UNITED STATES

Manner of Obtaining Composite Figures Used in Table 7

Strength as a beam or post[1]				Hardness[1]			
Values based on	Reduction factor[2]	Weight[3]		Values based on	Reduction factor[2]	Weight[3]	
		Green	Air-dry[8]			Green	Air-dry[8]
Static bending				Comp. perpendicu-			
M. of R.[4]	1.00	4	2	lar to grain	1.000	4	2
F. S. at E. L.[5]	1.80	2	1	End hardness	.865	2	1
Impact bending				Radial hardness	.930	2	1
F. S. at E. L.[5]	.80	2	1	Tangential			
Comp. parallel				hardness	.950	2	1
F. S. at E. L.[5]	2.80	2	1				
Max. cr. str.[6]	2.30	4	2				

Shock resisting ability[1]				Stiffness[1]			
Values based on	Reduction factor[2]	Weight[3]		Values based on	Reduction factor[2]	Weight[1]	
		Green	Air-dry[8]			Green	Air-dry[8]
Static bending				Static bending			
Work to max.load	1.000	4	2	M. of E.[7]	1.00	4	2
Total work	.380	2	1	Impact bending			
Impact bending				M. of E.[7]	1.00	2	1
Height of drop	.358	4	2	Comp. parallel			
				M. of E.[7]	1.00	2	1

Shrinkage[1]	
Values Based on	Weight[1]
Volume	2
Radial±Tangential	1

The per cent shrinkage in volume from green to oven-dry condition is based on volume when green.

[1]Formulæ showing relation of other properties shown in table to specific gravity (G):
Strength as a beam or post..20000 G
Hardness................... 4300 G[2].[5]
Shock resisting ability...... 44.5 G[2].[0]
Stiffness................... 3000 G
Shrinkage................. 26.5 G

[2]The reduction factor represents the average ratio of the first property, which is taken as unity, to the properties listed below as determined by the average of all species tested green.
[3]The weight taken into account, the relative importance of the various properties included in the composite values, and also the greater reliability of the values based on green tests due to the greater amount of data.
[4]M. of R.—Modulus of rupture.
[5]F. S. at E. L.—Fiber stress at elastic limit.
[6]Max. cr. str.—Maximum crushing strength.
[7]M. of E.—Modulus of elasticity.
[8]The air-dry values were reduced to 12 per cent moisture by the following approximate formulæ, which may be used within narrow limits:

When moisture is under 12 per cent

$$D12 - \frac{6(AD-B) + B}{18-M}$$

When moisture is above 12 per cent

$$D12 - \frac{10(AD-B) + B}{22-M}$$

D12—Value at 12 per cent moisture.
AD—Value air-dry as tested.
M—Per cent moisture as tested.

TABLE 7. SOME PHYSICAL AND MECHANICAL PROPERTIES OF BOX WOODS, ON THE BASIS OF WHITE PINE[1] AT 100

KIND OF WOOD	1 Specific gravity oven-dry wt. green volume	2 Shrinkage in volume from green to oven-dry	3 Strength as a beam or post	4 Hardness	5 Shock resisting ability	6 Stiffness
Coniferous species						
Pine, white (Pinus strobus)	100	100	100	100	100	100
	(0.363)	(7.9)	(7340)	(363)	(6.00)	(1235)
Northern white cedar	81	87	74	79	80	62
Cedar, incense	91	103	108	124	92	89
Cedar, Port Orford	113	147	128	143	157	140
Cedar, Western red	85	100	96	95	87	90
Cypress, bald	113	138	123	132	128	113
Douglas fir—Washington and Oregon (coast type)	125	161	140	154	135	151
Douglas fir—Montana and Wyoming (mount. type)	112	130	112	135	114	114
Fir, Alpine	84	117	82	92	62	77
Fir, Amabilis	103	180	103	96	117	118
Fir, balsam	92	130	87	79	85	93
Fir, lowland white	102	133	107	111	119	124
Fir, noble	96	173	106	97	118	123
Fir, white	96	130	103	116	93	106
Hemlock, Eastern	106	128	113	134	115	101
Hemlock, Western	110	151	121	124	108	121
Larch, Western	133	163	137	165	136	123
Pine, jack	108	129	95	122	136	88
Pine, loblolly	139	161	137	159	160	131
Pine, lodgepole	105	144	99	107	103	103
Pine, longleaf	152	157	164	198	175	151
Pine, Norway	121	147	128	125	143	132
Pine, pitch	129	151	112	150	162	103
Pine, shortleaf	136	146	138	169	157	125
Pine, sugar	99	106	98	107	87	89
Pine, Western white	108	146	110	98	121	123
Pine, Western yellow	105	127	98	109	97	94
Spruce, Englemann	86	129	82	86	77	82
Spruce, red, white, sitka	100	155	103	106	120	108
Tamarack	135	162	128	142	145	118

crates often fail on account of nail shear at the ends of the boards due to lack of shearing strength in the wood. No values for shearing strength are given in the tables in this book.

Strength as a Beam—The strength of a beam is its ability to support a load. The strength varies inversely as the length, directly as the width, and directly as the square of the height.

---

[1]The values in columns 2 to 6 are based on composite data derived as indicated in Table 6.

TABLE 7. SOME PHYSICAL AND MECHANICAL PROPERTIES OF BOX WOODS, ON THE BASIS OF WHITE PINE[1] AT 100—*Concluded*

KIND OF WOOD	1 Specific gravity oven-dry wt. green volume	2 Shrinkage in volume from green to oven-dry	3 Strength as a beam or post	4 Hardness	5 Shock resisting ability	6 Stiffness
Hardwood species						
Ash, black	126	182	104	164	203	98
Ash, pumpkin	134	143	121	270	151	94
Ash, white	144	152	146	264	239	124
Aspen	99	137	85	80	134	79
Basswood	90	200	87	78	91	100
Beech	150	205	137	238	216	120
Birch, paper	130	203	99	132	245	93
Birch, sweet	162	185	152	260	257	147
Birch, yellow	152	211	154	214	272	144
Buckeye, yellow	90	149	81	81	90	89
Butternut	99	127	89	105	137	91
Chestnut	109	141	97	130	120	89
Cottonwood, common	102	175	89	93	122	98
Cucumber tree	121	173	125	145	173	139
Elm, cork	158	173	145	269	317	115
Elm, slippery	134	175	128	186	272	111
Elm, white	120	180	115	151	199	98
Gum, black	127	168	114	197	141	94
Gum, cotton	125	154	119	204	143	102
Gum, red	122	190	118	156	170	113
Hackberry	134	175	104	194	249	87
Magnolia (evergreen)	127	154	109	209	252	108
Maple, red	134	156	132	199	177	125
Maple, silver	121	144	98	170	162	85
Maple, sugar	154	182	154	272	204	131
Oak, commercial white	162	196	137	291	214	120
Oak, commercial red	155	186	135	263	215	131
Poplar, yellow	102	143	99	100	94	116
Sycamore	126	173	107	169	133	103
Willow, black	96	160	61	92	165	55

For instance, a 10-foot beam is half as strong as one 5 feet long; a plank 8 inches wide is twice as strong as one 4 inches wide; a board 1 inch thick is four times as strong as one ½ inch thick, quality and the other two dimensions being the same in each case.

The computed stress in the outermost fibers of a beam at the maximum load is known as the *modulus of rupture*. The strength of these extreme fibers, per unit of cross-sectional area, varies in different species and is independent of the dimensions of the stick.

[1]The values in columns 2 to 6 are based on composite data derived as indicated in Table 6.

Stiffness—By stiffness is meant the resistance a beam offers to bending. It varies inversely as the cube of the length, directly as the width, and directly as the cube of the height. For instance, a 5-foot beam is eight times as stiff as one 10 feet long; a plank 8 inches wide is twice as stiff as one 4 inches wide; a board 1 inch thick is eight times as stiff as one ½ inch thick.

The *modulus of elasticity* is a measure of the comparative stiffness of beams of the same dimensions but of different species.

Shock-resisting Ability—Shock-resisting ability, often called "toughness," is important in box and crate material. The rough handling boxes receive makes it very desirable that box woods rank high in enduring shocks without breaking, although this property is often sacrificed for others more important commercially.

Hardness—By hardness is meant resistance, to indentation. It is important in boxes in that it indicates the ease with which nails may be over-driven and consequently influences the selection of nails with respect to size of head, and the ease with which label imprints may be made in the wood.

Nail-holding Power—By nail-holding power is meant the maximum resistance to be overcome in pulling nails out of wood. If the nails are driven into the side grain of the wood this resistance will be greater than if they are driven into the end grain. (See Table 10.)

CARE AND SEASONING OF LUMBER IN STORAGE[1]

It is usually necessary at box manufacturing plants to keep on hand a certain supply of lumber in excess of immediate demands. Such stock requires care to prevent deterioration and to promote seasoning as much as possible. Most of the seasoning, however, is usually done at the sawmill so as to avoid paying shipping charges on the excess moisture. For example, if wood containing 74 parts of moisture by weight per 100 parts of dry wood is dried down to 16 per cent moisture there have been removed 58 parts, or one-third of the total weight, and the freight charge is reduced correspondingly. Occasionally it is necessary to dry stock further at the factory.

---

[1]For a more detailed discussion of this subject see U. S. Department of Agriculture Bulletin 552, "The Seasoning of Wood," by H. S. Betts, 1917; and U. S. Department of Agriculture Bulletin 510, "Timber Storage Conditions in the Eastern and Southern States with Reference to Decay Problems," by C. J. Humphrey, 1917. These may be obtained from the Superintendent of Documents, Government Printing Office, Washington, D. C., at 10 cents and 20 cents, respectively.

POSSIBLE DETERIORATION IN STORED LUMBER

Checking at Ends and. on Surfaces—Although checking is always a possible cause of deterioration in stored lumber, the woods which are most commonly used for boxes and crates, namely, conifers and light hardwoods, do not check so badly as some of the heavier hardwoods.

Twisting and Cupping—Lumber which is not straight causes more or less trouble in manufacture and sets up stresses in the finished box when it is nailed down in a flat position. These difficulties can be largely avoided by proper piling of the lumber.

Casehardening, Honeycombing, and Collapse—Casehardening, honeycombing, and collapse do not develop seriously in the air-drying of most woods used for boxes. Oak, especially in the South, is apt to caseharden and honeycomb when exposed to summer atmospheric conditions.

Blue Stain or Sap Stain—Blue stain, or sap stain, is a blue discoloration of the sapwood. It is very common in the pines and red gum and occurs also in the sapwood of other species. Blue stain is due to a fungous growth, which lives on the sap in the cells, does not destroy the wood or injure its strength, and is objectionable only on account of the discoloration it produces. Badly stained pieces may make the presence of decay hard to detect

The fungus producing blue stain may occur in the log, but it occurs more commonly in freshly-sawed lumber. It can thrive only as long as the sapwood is moist; therefore, piling the lumber so that it will season as rapidly as possible greatly reduces, though it does not prevent, this discoloration. Blue stain makes rapid progress in green lumber during warm humid weather, especially when the lumber is close-piled, as it usually is in transit. Under such conditions the stain may penetrate all of the sapwood in a few days. Blue stain can be prevented by kiln-drying the lumber immediately after sawing; this is ordinarily done only with the higher grades, although some lumber mills also run the lower grades of lumber through a kiln. Another preventive measure consists in dipping the lumber as it comes from the saw in an antiseptic solution, such as sodium carbonate.

Decay or Rot—Decay is due to fungous growth which destroys the wood substance. In order that decay may take place, the wood must be moist and the temperature not too cold. Wood dried to below 20 per cent moisture rarely decays; therefore, box lumber dried to from 12 to 18 per cent moisture is practically immune from decay as long as it remains in that condition. Although decay is not so rapid in its action as sap stain, it may seriously reduce the strength

Fig. 8—Method of measuring twisting of plywood.

of some woods in 3 or 4 months during warm weather, especially when close-piled. Decay, including the so-called dry rot, can be prevented in stored lumber by properly piling the lumber some distance above the ground.

Insect Attack—Certain woods are subject to insect attack when insufficiently seasoned. The sapwood of some seasoned hardwoods is subject to attack by an insect known as the powder-post beetle. Hickory, ash, and oak are most subject to this injury, but butternut, maple, elm, poplar, sycamore, and others are also attacked. Containers made from such lumber should not be used in foreign trade because some countries will not allow such packages to enter for fear of the introduction of injurious insects.

Proper Methods of Piling Lumber in the Yard—The expense which it is advisable to incur in equipping a lumber yard for proper air seasoning of lumber depends largely on the permanency of its location. The small amount of additional work required for properly piling lumber so as to shorten the time required for seasoning and reduce deterioration is usually well worth while. Lumber thrown on the ground promiscuously, or piled on sagged foundations with loose projecting ends, will surely depreciate in value in a comparatively short time.

Fig. 9—Lumber piled sidewise on concrete and metal foundations.

A lumber yard should be well drained, and so situated and divided up by alleys as to reduce the cost of handling the lumber to a minimum.

Box lumber is practically always piled flat; it may lie with the ends of the boards toward the alley (endwise piling), or parallel with the alley (sidewise piling), as shown in figure 10. In either case the piles slope from front to rear, away from the alley. Endwise piling is more common because it facilitates handling of the lumber and because of the better visual inspection from the alley which it affords. Sidewise piling has the advantage of giving better air circulation from side to side, and what moisture enters the piles runs across the boards instead of running lengthwise and accumulating under the stickers as in end-piling.

FIG. 10—A well-kept lumber yard maintained by a large eastern wood-using factory. (Note forward pitch 'of stacks, treated ends, and general sanitary ground conditions.)

FIG. 11—Side view of lumber piled endwise to the alley with skids resting directly on the piers.

Foundations and Skids—Strong and durable foundations should be provided for the lumber piles. The best kind of foundation consists of piers of concrete or masonry, as shown in figures 9 and 11. If this form of construction is not feasible, creosoted wooden posts, or creosoted blocks, or supports of very durable woods may be used. Never use untreated sapwood or even heartwood of non-durable woods in the foundations except for very temporary purposes.

The tops of these foundations should be level in the direction parallel to the alley but sloping from front to rear 1 inch for every foot. The top of the lowest foundations should be sufficiently high so that, allowing for cross-pieces over the piers, the lumber will be at least 18 inches from the ground. Weeds and other obstructions to circulation should be removed from around the piles.

The distance between piers crosswise of the pile varies with the thickness of skids used, but should be such as to avoid any sagging in the skids.

The distance between piers parallel with the pile depends on whether the cross pieces, or skids, are laid directly on the piers, figure 10, or on beams placed on the piers parallel to the pile, figure 11. If the first method is used the distance between piers must be the same as between subsequent stickers, for the stickers must be aligned over the skids on the piers. This distance should not exceed 4 feet, and for lumber that warps easily, it must be less. If the last method is used in which the skids rest on strong beams laid on the piers parallel with the pile, fewer piers need be built; this method also permits changing the spacing of the skids and stickers for different kinds of lumber, and is especially recommended for red gum, black gum, and cotton gum (tupelo) for which it is best to have the stickers about 2 feet apart.

For the beams and skids steel I-beams or inverted railroad rails securely imbedded in the foundation are most permanent. Creosoted timbers, or naturally durable woods, are also very satisfactory. If the wood is given no preservative treatment, its life will be increased somewhat by applying two coats of hot creosote at all points of contact. Untreated sapwood so close to the ground will decay in a comparatively short time and may infect the lumber.

Stickers—The stickers should be of heartwood, preferably of some durable species, dressed on one side to uniform thickness and not over 4 inches wide. Narrower widths are recommended. It is very poor policy to use regular widths of lumber for cross pieces within the piles because little or no drying takes place where large areas are covered up, and decay may set in. The stickers should be $\frac{7}{8}$ inch thick for inch lumber and up to $1\frac{1}{2}$ inches for thicker stock; they should be slightly longer than the width of the pile.

The front and rear stickers should be flush with or protrude slightly beyond the front and rear of the lumber piles. The other stickers should be placed in alignment over the skids and parallel to the front of the lumber pile.

Placing of Lumber—If possible, different lengths of lumber should be put in separate piles. No loose and unsupported ends should be permitted. A space of about 1 inch should be left between the edges of 1-inch boards in each course, and 2 inches between 2-inch boards. Lumber piled in the open should have each course project slightly over the course beneath on the front side of the pile so as to provide a forward pitch to the high end of the stack. For wide piles it is recommended that a vertical open space or flue be left in the middle of the pile, about the width of a board, extending upward from the skid two-thirds the height of the pile.

The top of the lumber pile should be closed with overlapping boards laid so as to drain off all water. It is also desirable, especially for the better grades of lumber, to have this covering or roof project on all sides of the pile so as to keep out some of the snow and rain, and produce shade for the sides and ends.

Size and Spacing of Piles—Lumber piles are' usually built from 8 to 16 feet wide. The height depends on the character of the lumber, and the extent to which the yard is crowded. The space between piles should not be less than two feet; four or five feet is better if yard conditions permit.

Kiln-Drying Box Lumber[1]—Lumber 1 inch thick requires from 2 months to a year for air-drying, but the green stock can, as a rule, be kiln-dried for box purposes in from 2 to 10 days. Veneer or rotary-cut lumber $\frac{3}{16}$ inch thick requires from 6 to 12 days for air-drying; the same material can be kiln-dried in about 12 hours. Kiln-drying at the saw mill

---

[1]For information on the principles of kiln-drying and the operation of kilns see: Forestry Bulletin 104—The Principles of Drying Lumber at Atmospheric Pressure, and Humidity Diagram, by H. D. Tiemann; U. S. Department of Agriculture Bulletin 552—The Seasoning of Wood, by H. S. Betts; and U. S. Department of Agriculture Bulletin 509—The Theory of Drying and Its Application to the New Humidity Regulated and Recirculating Dry Kiln, by H. D. Tiemann; The Kiln Drying of Lumber, by H. D. Tiemann, J. B. Lippincott & Co., Philadelphia, Pa.

also prevents deterioration of the lumber, especially blue stain. The saving in time by kiln-drying greatly reduces the amount of stock it is necessary to carry in the yards. On the other hand, the cost of kiln equipment and the expense of kiln operation offset to some extent the advantages so gained.

In deciding whether it pays to kiln-dry lumber instead of air-drying it, the following factors should be considered·

Air-Drying	Kiln-Drying
Interest on capital invested in ground occupied by lumber yard, yard equipment, and the large amount of lumber kept in storage.	Interest on capital invested in ground occupied by kilns and trackage, dry kilns, and equipment, including extra boiler capacity, storage sheds, and a small amount of lumber kept in storage.
Taxes on land and equipment.	Taxes on land, kilns, sheds, and equipment.
Insurance on equipment and lumber.	Insurance on buildings, equipment and a comparatively small amount of lumber.
Depreciation of lumber and equipment.	Depreciation of buildings, equipment and lumber.
Cost of handling lumber from the time it is received until ready for shipment or manufacture.	Cost of handling lumber from the time it is received until ready for shipment or manufacture.
	Cost of operating of kilns: attendance, fuel, water, etc.

THE USE OF VENEER IN THE CONSTRUCTION OF PACKING BOXES

Definition—Veneer is a thin sheet of wood. There is no standard thickness above which it is called lumber. The common practice is to use the term veneer for all stock which has been cut on special veneer machinery, and lumber for that which is cut with ordinary circular or band saws.

Although originally veneer was cut from high-priced cabinet woods to save material, it is now cut extensively from common species, and is used for many purposes when lightness rather than beauty is the principal requisite. It is well suited for the manufacture of small packages and even packing boxes of considerable size because of its light weight and the small amount of wood required for construction of this kind.

MANUFACTURE OF VENEER

Method of Cutting—Most of the veneer used at present in box manufacture is resawed, rotary-cut, or sliced. The rotary and sliced methods reduce the waste and produce wide stock. The veneer thus produced is comparatively free from defects, as only relatively smooth logs can be used for this purpose. Such material is cut with a special thin-edged veneer saw, a knife, or an ordinary band saw.

Drying—Veneer, like other forms of lumber, should be properly dried in order to give satisfactory service. Drying is sometimes done in the open air in open sheds, the time required varying from several days to several weeks, depending on the kind and thickness of the stock and weather conditions. This thin material may be dried in a kiln, in which case it must be weighted down to keep it straight. It is often put through progressive driers which dry the wood in from several minutes to an hour. These driers consist of long chambers with a series of belts or live rolls on which the veneer is carried through the apparatus. The temperature is comparatively high and the humidity low, so that rapid drying results. There is little danger of casehardening very thin stock. Another type of drier consists of a series of heated iron plates between which the material is pressed. These plates separate at regular intervals so as to allow the material to shrink.

Woods Used for Box Veneers

Thin lumber is made from many kinds of woods, including most of our commercial species, but red gum leads all others in quantity used. Yellow pine and maple are also used largely in the manufacture of boxes, baskets, and crates. Cottonwood is a very desirable species for use in the production of veneer because it gives very little trouble in cutting. The sides and bottoms of cracker boxes and light egg cases are made principally of cottonwood stock. Elm is used largely for cheese boxes. Other species commonly used for thin stock in box and package manufacture are birch, beech, cotton gum (tupelo), basswood, sycamore, and Douglas fir; and many other commercial species are used to a smaller extent.

Although very definite statistics are not available as to the amount of thin lumber used for boxes and fruit and vegetable packages, it is estimated that out of the total 6 billion board feet of lumber annually used by the box industry, 700 million feet log scale is used for thin stock. The use of thin lumber is gradually increasing.

Most of the thin lumber or veneer used in boxes and

crates is in single thicknesses securely fastened to relatively thick ends or cleats.

Examples of the use of single-thickness stock for shipping containers are very common. The cheese box is one of the oldest forms. Vegetable barrels, baskets, berry boxes, fruit crates, cracker boxes, egg cases, canned goods boxes, and many others, including a special type known as wirebound boxes, are used extensively.

Use of Plywood **in** Packing Boxes—The properties of veneer or thin lumber in single thicknesses are improved by gluing together three or more sheets with the grain crossing. The product is known as "plywood," and has the advantage of producing a comparatively strong and light piece of material in which the strength, stiffness and shrinkage along and across the grain of the face pieces are more nearly equal than in lumber. Plywood also has greater resistance to splitting in nailing and to puncturing in handling.

Plywood is not used very extensively in the manufacture of boxes, but it has distinct advantages over other forms of wood construction. A box properly made of plywood is exceedingly strong for its weight. The principal objection to the more extended use of plywood in boxes is the cost involved in gluing up of the thin sheets.

Plywood should be made up of an odd number of plies with the grain of successive plies at right angles to each other. The construction on the two sides of the core should be symmetrical as to species, thickness of panels, and direction of grain. The strength in bending is less in the direction parallel to the grain of the face pieces, and greater at right angles to the grain of the face pieces than in boards of the same thickness and same kind of wood. A combination of faces of strong wood and a thick core of a light wood gives greater strength in bending than the same faces with a thinner core of the same weight of some heavier species. The glued surfaces are not so likely to separate in plywood constructed with thin plies as in that made of thicker material.

Plywood does not split in nailing so easily nor does it puncture so easily as a single board of the same thickness, the resistance increasing with the number of plies. The greater the number of plies, the straighter the plywood will remain with change in moisture content. Plywood in outdoor service will cup and twist least if it has a moisture content of from 10 to 15 per cent when it comes from the glue press.

CHAPTER II

BOX DESIGN

Factors Influencing Details of Design—Characteristics of
the Various Styles of Boxes—Factors Determining the
Amount of Strength Required—Factors Determining
the Size of a Box—Special Constructions.

Box design may be defined as the development of definite
details for constructing boxes which will deliver their contents to the purchaser in a satisfactory condition and at a
minimum cost. The construction of more expensive boxes
can not be justified unless they are to serve as an advertise-
ing medium or perform some other service which warrants
the additional cost. Among the factors which affect box
economy is the cost of the following: raw materials, manu-

TABLE 8. Thicknesses of Box Boards Obtained by Resawing or Dress-
ing 4/4 to 7/4-inch Lumber[1]

Box boards, thickness Inches	Rough lumber, thickness[2] Inches
3/16 rough or S1S	3 pieces from 4/4
1/4 rough or S1S	3 pieces from 4/4
5/16 rough or S1S	3 pieces from 5/4
3/8 rough or S1S	2 pieces from 4/4
7/16 rough or S1S	2 pieces from 4/4
1/2 rough or S1S	2 pieces from 5/4
9/16 rough or S1S	2 pieces from 5/4
5/8 rough or S1S	2 pieces from 6/4
11/16 rough	2 pieces from 6/4
11/16 S1S	2 pieces from 7/4
3/4 rough or S1S	2 pieces from 7/4
13/16 rough, S1S, or S2S	1 piece from 4/4
7/8 rough, S1S, or S2S	1 piece from 4/4
15/16 rough or S1S	1 piece from 4/4
4/4 rough	1 piece from 4/4
4/4 S1S	1 piece from 5/4

If box parts are to be dressed on two sides (S-2-S) and to be full thickness also, use next thickness of lumber except where specified in above table.

[1]Adopted by the National Association of Box Manufacturers—August 5, 1915
[2]If full thickness without variation is required, then use the next greater
thickness.

facturing (including assembling), handling, storage, freight, and losses due to box failures. A designer of boxes should endeavor to gain a knowledge of all these phases of the problem.

FACTORS INFLUENCING DETAILS OF DESIGN

Lumber and Veneer

Availability and Supply—Lumber of suitable thickness for box construction is usually obtained by resawing regular sizes of low grade stock, although in some sections the logs are sawed directly into lumber of the desired thicknesses. The designer should have definite information regarding the properties, grades, widths, thicknesses, lengths, supply, and cost of lumber of the species available where the box is to be manufactured and the commodity for which the box is made, and the manner in which the commodity is to be packed. Data showing what thicknesses can be obtained from commercial lumber by resawing and surfacing should be at hand. (See Tables 8 and 9.) The dimensions of mate-

TABLE 9. STANDARD THICKNESSES OF HARDWOODS

Adopted by the National Hardwood Lumber Association and the American Hardwood Manufacturers' Association.[1]

Rough thickness Inches	Surfaced thickness Inches
3/8 S-2-S t	3/16
1/2 S-2-S t	5/16
5/8 S-2-S t	7/16
3/4 S-2-S t	9/16
1 S-2-S t	13/16
1 1/4 S-2-S t	1 3/32
1 1/2 S-2-S to	1 11/32
1 3/4 S-2-S to	1 1/2
2 S-2-S to	1 3/4
2 1/2 S-2-S to	2 1/4
3 S-2-S to	2 3/4
3 1/2 S-2-S to	3 1/4
4 S-2-S to	3 3/4

Lumber surfaced on one side only must be 1/16 inch full of the above thickness.

rial in boxes made of sawed lumber must, if not inconsistent with other requirements, be such as will use the material with the least waste in resawing, surfacing and cutting to size.

[1]Formerly the Hardwood Manufacturers' Association.

The main waste is due to trimming to required length and width and at the same time eliminating checked ends, splits, loose and rotten knots, and knot holes. Rotary-cut veneer can be obtained in lengths up to 60 inches and in any width that can be readily handled.

Cost—The actual cost of lumber is not the purchase price, but the cost of the usable material after due credit has been allowed for salvage of waste. Obviously it is important that the amount of waste which occurs in the various grades of material be accurately known to the designer.

Manufacturing Limitations

Equipment—Design will necessarily be influenced by the box-making equipment available. Equipment for making the common types of boxes is more or less standardized. Accurate information on the kind and cost of each operation performed and the quantity of work produced by each machine is therefore requisite to the design of such a box as will cost the least to manufacture.

Details which require the installing of special machinery for their execution should be avoided if possible, unless they are improvements of such a character that there will be a continued demand for their application. The experience of the Forest Products Laboratory, however, is that in a large majority of cases the common designs of boxes are the more efficient and that the special features in other designs usually interfere with balanced construction.

Cost of Operation—The cost of various machine operations depends on numerous factors such as general factory overhead charges, depreciation on machines, power charges, cost of tools, operator's wages, and volume of work done. The standardization of box and crate design has proved one of the chief factors in the reduction of cost of manufacture. Unusual styles or special features always increase the cost and as a rule decrease the serviceability of shipping containers.

Styles of Boxes—There are a number of styles of nailed wooden boxes so universally used that they may be called standard nailed boxes. (See Plate III.)

Many special styles of boxes have been developed for particular conditions and commodities. Whether or not a box which can be returned and refilled should be adopted for carrying a commodity depends wholly upon the economic phases of the problem. If box materials continue to increase in cost, the use of returnable boxes will no doubt increase. Returnable boxes should, so far as possible, be made collapsible, as they will then occupy less space in shipment and storage when empty.

Under some conditions boxes which can be easily and quickly opened are demanded. This requirement has been especially urgent in many of the United States Army Ordnance boxes. Types of such easily-opened boxes are shown in Plate IV.

BALANCED CONSTRUCTION AND FACTORS AFFECTING STRENGTH[1]

When all elements in the construction of a box resist equally the destructive hazards of service, it is balanced in construction. A box may be balanced in construction and yet be excessively heavy, too strong, and uneconomical in the use of material; or it may be too light and weak for service. With unbalanced boxes which render satisfactory service there is frequently a waste of material in the stronger parts; and an equally or even more serviceable box may be obtained by reducing the strength of the stronger parts until they are in balance with the weaker parts. This is because the parts which are excessively heavy transmit an undue amount of the shocks and stresses to the lighter parts, thus causing the lighter parts to fail sooner. With a balanced box there is a more even distribution and absorption of stresses and shocks. Excessive thickness of lumber in sides, top, or bottom of a box will also produce undue stresses in the nails and, under certain conditions, will be a source of weakness.

The chief problem in box design is to detail the parts so that balanced construction and proper strength are both obtained, and at minimum cost. Balanced construction and a proper degree of strength can be determined by suitable methods of testing.[2]

Width of Stock and Joints—Stock which is wide enough to make one-piece box parts has various advantages in comparison with narrower stock. Joints which would otherwise occur are avoided, thus increasing the rigidity of the boxes and their resistance to "weaving action." In boxes of Style 1,

---

[1]For strength data on various types of boxes, see Forest Service Circular No. 214, obtainable from the Superintendent of Documents, Washington, D. C., at five cents a copy.
[2]See chapter IV, page 87.

Plate III, having no cleats but single-piece ends or two-piece with corrugated fasteners, it is very desirable to have one-piece sides because they diminish the liability to failure and make a more dependable construction. A tighter box is insured with single-piece parts. Larger knots can be permitted in boxes with single piece parts because the allowable sizes of knots bear a direct ratio to the width of the board. Single-piece parts reduce machine and labor costs; but if required exclusively, they would greatly increase material costs.

When two or more pieces are used in any part, the abutting edges are usually tightly joined. This may be aecom plished in several ways; the simplest form of joint between the edges of two boards is shown in figure 1, Plate XV. The abutting edges of the pieces should be straight and square with their faces, and in contact throughout so as to make a tight joint.

When some other than the butt joint is used for joining edges of boards, it is called a matched joint. One type of matching which is used to a limited extent in box construction when the lumber is $\frac{5}{8}$ inch or more in thickness is shown by figure 6, Plate XV. Such lumber, known as shiplap, milled to join in this way, can be obtained in various widths, and this is undoubtedly the reason why it sometimes appears in box construction. It is not as effective for tight construction as other matched joints since the adjacent boards can bend independently.

The matched joint illustrated by figure 5, Plate XV, is very commonly used for joining box boards. In addition to matching, the pieces may be glued or fastened together with corrugated fasteners. This construction makes it possible, when the material is over $\frac{5}{8}$ inch thick, for the weaker boards and the boards which receive the more severe thrusts in service to be supported to some extent along their edges by the adjoining boards.

An excellent matched joint (Linderman) for box work is shown in figure 3, Plate XV. The taper lengthwise in this joint produces a wedging action between the parts and binds the two pieces tightly together when they are forced into proper position. As the two pieces are forced together, glue is applied to the uniting surfaces, which, if the gluing is properly done, increases the strength of the joint and enables the combined parts to approximate a single piece in character; it is less effective in material ½ inch or less in thickness.

In constructing boxes such as Styles 1 and 6, Plate III, and those in figures 1 and 2, Plate IV, joints in the sides and ends should be so located that there is considerable distance between their respective planes, to avoid a line of weakness around the box.

An important advantage of all matched joints is that a tight box is maintained even though some shrinkage occurs.

Corrugated Fasteners—In figure 2, Plate XV, several types of fasteners are shown. The fasteners with parallel corrugations may also be obtained in continuous coils (figure 4, Plate XV) for use on automatic driving machines which cut and drive the fasteners in one operation.

These fasteners may be used for holding pieces together and preventing relative endwise movement in the joints (figures 1, 5, and 6, Plate XV) and for preventing relative endwise movement of pieces in Linderman joints when poorly glued (figure 3, Plate XV). They are also driven across splits and large checks to prevent further development of such defects; greater efficiency is obtained in such cases if the fasteners are driven from both sides.

The depth of the fasteners should be slightly less than the thickness of the pieces into which they are driven so that the cutting edge may not protrude after driving. The fasteners with divergent corrugations have a tendency to draw the pieces closer together when they are driven.

Physical Properties of Wood—The physical properties of wood have a vital influence on the strength of a box. The effect of using material the density of which is much lower than the average for the species is to produce a box in which the low density parts of it are weak and will usually break quickly in service.

Material of low bending resistance is not suitable for long boxes in which the contents are of such a nature that considerable stiffness in the material is required to maintain the shape of the box.

Increased resistance to splitting[1] is desirable, especially for ends which are not cleated or otherwise reinforced. If some method of reinforcing against splitting must be used, the cost of the box is increased unless the extra charges are

---

[1]See also nailing qualities of wood, page 51

balanced by a reduction in the thickness of material made possible by such reinforcement.

Failures by puncturing are infrequent. Inspectors at the ports of embarkation during war shipments estimated that less than one per cent of boxes inspected showed damage due to puncturing. The damage due to this cause when thin mate-

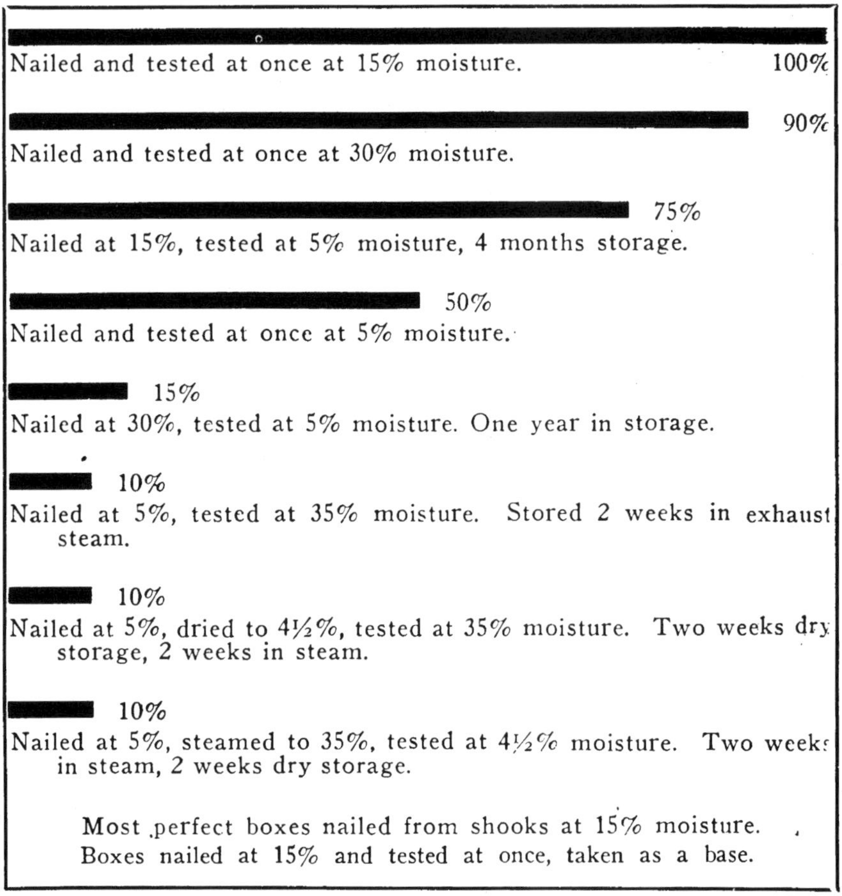

Nailed and tested at once at 15% moisture.   100%

                                                              90%
Nailed and tested at once at 30% moisture.

                                              75%
Nailed at 15%, tested at 5% moisture, 4 months storage.

                              50%
Nailed and tested at once at 5% moisture.

        15%
Nailed at 30%, tested at 5% moisture. One year in storage.

      10%
Nailed at 5%, tested at 35% moisture. Stored 2 weeks in exhaust steam.

      10%
Nailed at 5%, dried to 4½%, tested at 35% moisture. Two weeks dry storage, 2 weeks in steam.

      10%
Nailed at 5%, steamed to 35%, tested at 4½% moisture. Two weeks in steam, 2 weeks dry storage.

Most perfect boxes nailed from shooks at 15% moisture.
Boxes nailed at 15% and tested at once, taken as a base.

Fig. 12—Effect of condition and change of condition of lumber on strength of boxes in storage. Boxes for 2 doz. No. 3 cans, nailed with seven cement-coated nails to each nailing edge. Chart based on tests to date. Data insufficient for accurate comparisons.

rial must be used can be reduced by substituting plywood for veneer, since it offers more resistance to puncturing. If warped lumber or veneer is used, initial stresses are produced when such material is forced to assume the proper shape in assembling the box. Such initial stresses decrease the amount of strength remaining for resisting the hazards of service. Boxes made of such material may also be somewhat misshapen.

Moisture Content[1]—The moisture content of the material used in constructing boxes influences their strength greatly, and in various ways. An abnormal amount of moisture in box material causes the points of the nails to pull more easily from the wood, the heads to pull through the wood, and the shanks of the nails to shear out at the ends of the boards. When boxes made of green or wet material subsequently dry out, the nails become loose and pull easily. The boards also split and check because the nails resist the shrinkage, which is the normal result of drying. Variations and changes in the

FIG. 13— Effect of shrinkage on strapped boxes. Boxes made and strapped at a 30 per cent moisture content; the boxes were photographed after drying out to 10 per cent moisture content.

amounts of moisture contained in box lumber affect the strength of a box, as indicated by the results of tests given in figure 12.

The moisture content of box material has a very great influence on the maintenance of the strengthening effect of strapping and wire bands. If shrinkage occurs after strapping is nailed on, the strapping buckles between the nails. (See figure 13.) Straps with ends joined by some sealing or

---

[1]See page 15 for general discussion of moisture content.

clamping device often become so loose through shrinkage of the wood that unless nailed in place they will easily slip off the box. Inspectors have reported that large quantities of straps which have slipped off boxes, with the seals unbroken, are at times left in freight cars after unloading. A similar loosening effect may occur when boxes are stored for a considerable time. The effect of shrinkage on the usefulness of wire bands is similar to that on strapping.

It is apparent that when straps or wire ties are to be used, the box material should have such a percentage of moisture as it is likely to retain after construction and packing is completed; this is usually between 12 and 18 per cent. Strapping should, if possible, be put on boxes under maximum tension immediately prior to shipment, as the bad buckling and loosening effects from shrinkage in storage are not then so apt to occur.

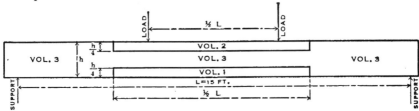

Fig. 14—Division of a beam into volumes for describing the location of knots.

Defects[1]—Boxes are ordinarily made of low-grade lumber containing various defects. Such defects as do not affect the serviceability of the box should be allowed to remain, in order that the amount of waste, and consequently the actual cost of material, may be the minimum.

Larger knots may be permitted in wide pieces than in narrow widths, but knots should not be permitted to interfere with proper nailing. A knot hole, besides being weakening, may cause loss of the contents of a box. Knots are most weakening in that part of a beam indicated by volume 1 in figure 14, next if located in volume 2, and least weakening if on any part of volume 3. Figure 15 shows a crate slat which failed on account of too large a knot at A.

Knots should be limited as to size, because the weakening effect is practically proportional to their effective diameter, measured as shown in Plate I. A satisfactory method of lim-

---

[1]See pages 8 to 10 for description of various defects.

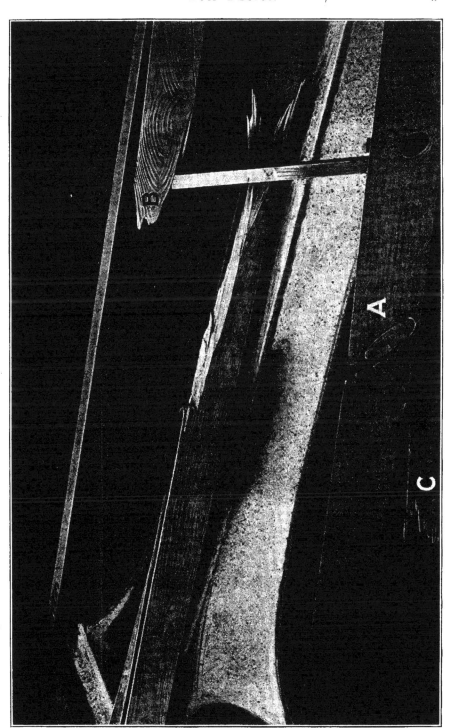

FIG. 15—Broken crating of 1 mental concrete lighting post. A. Knot. B. Cross grain. C. Decay.

iting the size is to require that the effective diameter shall not exceed a certain fraction of the width of the board.

Shakes and checks are objectionable in material for boxes for certain commodities. They may develop into splits and

TABLE 10. HOLDING POWER OF NAILS IN SIDE AND END GRAIN OF VARIOUS SPECIES

7d Cement-coated nails driven to a depth of one inch and pulled immediately

Species	Per cent of moisture	Specific gravity	Withdrawing pull in lbs.	
			End grain	Side grain
Group I[1]				
Pine, white	7.7	.391	122	203
Pine, Norway	7.4	.507	149	254
Pine, jack	7.6	.429	145	245
Aspen	6.5	.412	141	186
Spruce, red	10.7	.413	133	199
Spruce, white	7.6	.396	131	196
Pine, Western yellow	7.2	.433	96	196
Cottonwood	6.8	.343	129	177
Basswood	6.5	.412	124	175
Fir, white	7.6	.437	101	183
Cedar	9.3	.315	93	144
Group II				
Hemlock	8.6	.501	139	236
Pine, Southern loblolly yellow	7.7	.516	142	268
Longleaf	8.2	.599	196	313
Group III				
Elm, white	8.2	.537	212	305
Gum, heartwood	6.0	.488	179	243
Gum, sapwood	8.1	.433	189	220
Sycamore	7.0	.552	243	314
Maple, silver	6.8	.506	252	304
Group IV				
Maple	9.3	.643	350	406
Ash, white	8.9	.640	347	407
Beech	8.4	.669	322	414
Oak, cow	4.3	.756	277	323
Oak, post	7.3	.732	351	345
Oak, red	7.6	.660	297	333
Oak, white	7.3	.696	268	289
Birch	8.6	.661	298	406

cracks and thus increase the liability of the box to fail in service. A board containing large checks or shakes which extend through from one face to the other should be considered as two boards. One method of preventing splits, checks, or shakes from increasing in size is to drive corrugated fas-

---

[1]Data are not available on all woods in each group.

teners across the apparent line of development. These corrugated fasteners should be driven only when the board can be firmly supported opposite the point of driving, as otherwise the attempted remedy may prove to be detrimental.

Cross grain, a slope of the fibers with respect to the main axis of a stick, is one of the most serious defects affecting the strength of box and crate material because it is very common and not easily detected.

Cross grain in box lumber is detrimental because it increases the danger of failure in boards which are subjected to bending and puncturing stresses; also because it makes the wood more susceptible to splitting when nails are driven where the defect occurs.

Insect holes, if large and present in sufficient numbers, frequently impair the strength of box lumber. They also affect the appearance and tightness of boxes, but in some cases such material can be used with a resulting saving in cost.

Rot is often found in low grades of lumber; the extent to which it may be permitted in box and crate material depends on the purpose for which the container is intended. Rot should not be allowed in pieces subjected to great stresses, or wherever nails are driven. The slat C at the bottom of the crate in figure 15 broke because it was partly decayed.

Occasional worm holes in wood do not seriously weaken it, but pieces which are badly perforated should not be used where strength or nail-holding power is essential. As a rule, worm holes indicate decay in the material in which they occur.

Nailing Qualities of Wood—The nailing qualities of the wood are of vital importance in box construction. It is wasteful practice to make a nailed box of lumber of considerable strength unless the parts are nailed together in such a way as to balance the construction.

The serviceability[1] of a nailed joint varies with the density of the wood, the ease with which it is split and sheared by nails, the initial moisture content, changes in moisture content[2], the character and location of defects, and the direction of the nails relative to the grain of the wood.

It will be observed from the preceding table that, in general, the difference between the resistances of the nails to

---

[1]See page 53 for the effect of nails on the strength of a joint.
[2]See page 15, moisture content.

Load in Pounds Required to Pull One Nail.

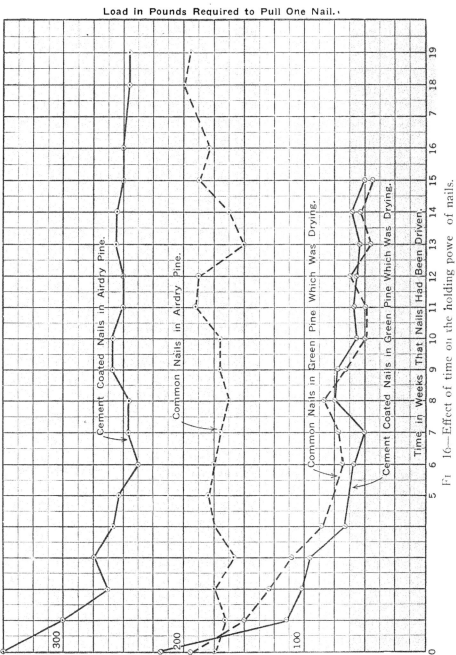

Fɪ 16—Effect of time on the holding powe of nails.

pulling from the end grain and from the side grain is greater for the light soft wood than for the heavier dense ones.

At times it is necessary to use denser woods for the cleats or ends or both than are used for the other parts of the box in order to secure sufficient nail-holding power to balance the construction.

Some woods are very susceptible to nail splitting; also nails easily shear out at the ends of the boards of some species. Both of these difficulties increase the probability of the ends being pulled away from sides, top, and bottom of the box. Nails driven in checks, rot, etc., have little holding power. The holding power of nails changes with the lapse of time after driving, as shown in figure 16.

Tests on the holding power of nails driven parallel with and at an angle to the grain, as shown in figure 3, Plate VII, show no appreciable difference in holding power when pulled immediately. In these tests the direction of pull was perpendicular to the surface of the board. Nails were similarly driven in green pine, which was then dried thoroughly in an oven at a temperature of 100° C, before testing; and it was found that while the diagonally-nailed pieces started to fail at lower loads than the straight nailed, the diagonally-nailed pieces soon developed more strength with the result that the force required to withdraw the nails was considerably higher than for those with straight nailing. These tests indicate that diagonal nailing may be of some advantage if boxes are to remain in storage where the moisture content will be reduced.

Fastenings and Reinforcements—Nails are the most common fastenings for box materials. The serviceability of a nailed joint varies with different nail characteristics and details of nailing, such as the character of the shank surface, the length and diameter of the shank, the flexibility of the shank, the size of the head, and the number or spacing of the nails. A special type of nails known as box nails[1] is made for use in the box industry. In order to minimize the splitting of material these nails are made of smaller wire than the ordinary plain wire nails used in building construction, though they must be of sufficient diameter not to bend or kink in driving. Another advantage of slender nails is that they are not so readily loosened in the wood as those of larger diameter and equal length, because the slender nails bend more readily un-

[1]See pages 139-140 for table of sizes of nails, etc.

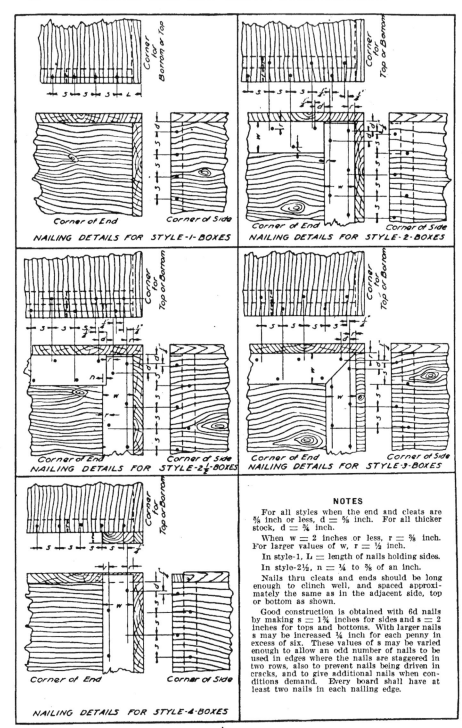

NOTES

For all styles when the end and cleats are ⅝ inch or less, d = ⅝ inch. For all thicker stock, d = ¾ inch.

When w = 2 inches or less, r = ⅜ inch. For larger values of w, r = ½ inch.

In style-1, L = length of nails holding sides.

In style-2½, n = ¼ to ⅜ of an inch.

Nails thru cleats and ends should be long enough to clinch well, and spaced approximately the same as in the adjacent side, top or bottom as shown.

Good construction is obtained with 6d nails by making s = 1¾ inches for sides and s = 2 inches for tops and bottoms. With larger nails s may be increased ¼ inch for each penny in excess of six. These values of s may be varied enough to allow an odd number of nails to be used in edges where the nails are staggered in two rows, also to prevent nails being driven in cracks, and to give additional nails when conditions demand. Every board shall have at least two nails in each nailing edge.

FIG. 17—Details for nailing standard styles of boxes for domestic shipment.

der the shocks of rough handling and the weaving strains that a box receives in transportation and are not worked back and forth their full length. If nails are too slim, however, the excessive bending, which readily occurs, will frequently cause them to break between the parts which they unite. Box nails may be obtained cement-coated, or with plain or barbed shanks. The cement coating increases the friction between the shank of the nail and the wood. Barbed nails are so called because the shanks of the nails have on their surfaces a series of small barbs or teeth.

The holding power of nails of the same kind but of different sizes against withdrawal in line with the shank is for ordinary sizes approximately proportional to the amount of shank surface in contact with the wood. If more holding power is needed, it is preferable to increase the number or length of the nails rather than their diameter. In this way the advantages of slim nails are retained, and the additional shank surface is secured with less additional metal in the nails.

The size and spacing of nails should be such as will not cause an unreasonable number of failures because of splitting the material in driving.

One of the difficult problems in a nailed box is to secure sufficient nail-holding power to stress all the wooden parts of a box to their maximum, and still maintain balanced construction. In some species of wood a large number of small nails may be required, and in others a smaller number of larger nails may be necessary to secure the desired results. Box woods are divided into four groups according to their nailing qualities.[1]. The nailing schedule on page 102 gives information for proper nailing of boxes constructed of woods from these groups.

Directions for Nailing—General directions and details for nailing cleats, sides, top, and bottom to ends of several styles of boxes for domestic shipment are given on page 103. (See also figure 17.) For foreign shipment the spacing should be about one-half inch less than given on page 102.

In figure 18 are shown the relative amounts of rough handling required to cause loss of contents in boxes constructed with various spacing of nails. A box with seven nails per nailing edge is taken as the basis for comparison.

The size and spacing of nails in some instances depends largely on the properties of the material upon which the heads

[1]See page 100 for grouping of species.

of the nails rest. If this material is of low density and easily sheared by the nails, it is advisable to have them closer to-

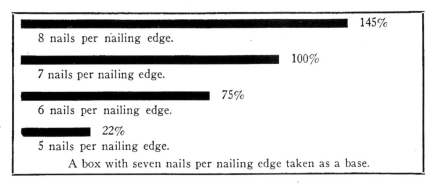

Fig. 18—Relation of number of nails to amount of rough handling required to cause loss of contents. Nailed boxes for 2 doz. No. 3 cans.

gether than would be required with a denser wood offering more resistance to such shearing action.

Placing nails closer together gives more holding power per inch of nailing edge for the wood in which the points are held, unless the nails become so close that their holding power is reduced by the splitting of the wood.

Side Nailing—The term "side nailing" refers to the nailing of the top and bottom to the edges of the sides. If the boards are less than $\frac{7}{16}$ inch in thickness, such nailing will add little to the serviceability[1] of the box but will make a tighter box, provided the strains due to the contents and the hazards encountered are not severe enough to spring the boards and produce nail splitting at these edges. The size of nails should be as shown by the nailing schedule on page 102. Six penny nails and smaller should be spaced approximately six inches apart, and this distance increased one inch for each penny over six.

Nails for Clinching—The character of the surface of the shank is relatively unimportant in nails to be clinched, the head and the clinched end being the important factors. Slender nails are to be preferred because they lessen the danger of splitting the wood and clinch more easily. The use of cleating nails made with a "side point" is increasing. The long slender point turns back into the wood when driven and gives more uniform results than right-angled clinching. .

---

[1]See discussion of strapping, page 60.

Large Nail Heads[1]—Tests have demonstrated that large nail heads have considerable advantage over small ones be-

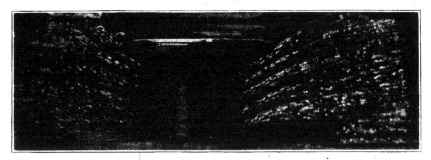

Fig. 19—Injury to wood fiber resulting from overdriving nails.

cause of the additional resistance offered to being pulled directly through the wood. Large heads also prevent the shanks of the nails from shearing out as easily at the ends of the

TABLE 11. EFFECT OF SIZE OF HEADS ON STRENGTH OF A NAILED JOINT

Material	Size of nail	Diameter of heads inches	Tension			Shear		
			Load per nail pounds	Heads off % total	Nails broken % total	Load per nail pounds	Heads off % total	Nails broken % total
5/16-inch red gum rotary cut	6d	.36	312	0	0	186	0	0
	6d	.26	257	0		155	0	0
	6d	.24	250	0		154	0	0
	5d	.28	213	100		158	33	16
	5d	.22	196	22	0	152	8	8
	5d	.19	212	0		164	0	0
	4d	.28	239	83		173	7	16
	4d	.22	248	100		168	63	0
	4d	.19	222	0		163	0	0
3/8-inch sawed white pine	6d	.36	220	11	0	164	0	0
	6d	.26	194	0		139	0	0
	6d	.24	181	0		139	0	0
	5d	.28	178	33		138	33	17
	5d	.22	149	0		142	8	8
	5d	.19	122	0	0	134	0	0
	4d	.28	154	22		172	33	67
	4d	.22	146	0		161	11	22
	4d	.19	116	0		150	0	0

boards when thin material is used. In many nails with extra large heads the material where the shank joins the head is so thin that failures frequently occur in the nail at this point, thus preventing to a large extent any addition to the strength of the box. Large headed nails are advantageous for mate-

[1]See page 139 Appendix for table of size and thickness of heads.

rial which contains many checks or which splits easily and for thin and low-density material; in fact, when thin material is being nailed the size of the head and the length of the nail are of nearly equal importance, with the necessity for large

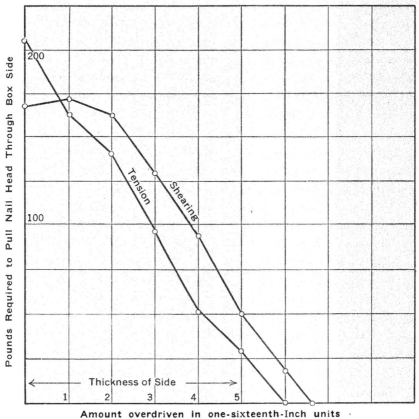

Amount overdriven in one-sixteenth-inch units

Fig. 20.—Effect of overdriving nails

heads increasing as the thickness of the material is reduced.

In Table 11 are shown results on approximately 100 nails with various sizes of heads. The small heads are the standard sizes for the different nails. It is shown that a very large percentage of the heads pulled from the four or five penny nails.

Overdriving Nails—One of the serious. faults in nailing boxes is overdriving. Nails should be driven until the top of the head is just flush with the surface of the material. Overdriving crushes and injures the wood fibers (figure 19) and decreases the strength to an extent which depends on the amount of overdriving (figure 20). On the other hand, if the heads of the nails are not driven flush with the surface, the joint between the boards is not so tight and rigid as it would be otherwise, also there is danger of the nail heads catching on objects, especially strapping on other boxes.

Screws—Screws are an admirable fastening when properly driven. They permit a box to be easily opened without danger of injury to box or contents from bars, chisels, or other tools. Boxes made with screws are also easily closed again for reshipping, which feature may be objectionable with reference to theft of goods in transit. Some method of sealing boxes which are closed with screws may be used to lessen the thieving losses. Other objections to screws are their cost, the cost of proper driving, and the great tendency to drive screws improperly their full length with a hammer, thereby sacrificing much of the strength that they should give.

The data[1] in Table 12 show the effect of different methods of driving on the holding power of No. 12 screws.

TABLE 12. RESISTANCE TO WITHDRAWAL OF NO. 12 SCREWS

Screws 1¼ inches long, driven to one inch depth in holes ⅛·inch in diameter.

Kind of wood	Method of driving			
	Screw driver		Hammer[2]	
	Pounds	Per cent	Pounds	Per cent
Basswood...................	478	100	281	59
Yellow pine	1144	100	681	60
Red gum	687	100	403	59
Birch......................	841	100	570	68

Staples—Staples are used for fastening both wire bands[3] and flat metal straps in place. To secure good holding power they should be driven for a considerable distance into solid wood and if driven into thin material the points should be clinched.

In holding box-strapping in place, staples have an advantage because they do not weaken it by puncturing as nails do, and they also hold the edges of a strap down together and present a curved part to strike against the edges of straps on another box, thus lessening the danger of catching and tearing them off.

[1]See page 63 for information on cleats fastened with nails, wood screws, and drive screws.
[2]One final turn with screw driver.
[3]See discussion of wirebound boxes on page 67.

Staples will not, as nails do, hold the tension in strapping, and therefore some method must be used of accomplishing this by sealing or fastening the ends together.

Strapping and Wire Bindings—The use of metal bindings around boxes may serve one or more of the following purposes:

1. To reinforce the package and increase its serviceability.

2. To minimize pilfering.

3. To secure lighter, yet equally serviceable packages.

Types of Metal Bindings

$$
\text{Metal bindings}\ldots
\begin{cases}
\text{Flat straps}\ldots
\begin{cases}
\text{Annealed} \\
\text{Unannealed}
\end{cases} \\
\text{Wires}\ldots\ldots
\begin{cases}
\text{Single wire} \\
\text{Two or more wires twisted}
\end{cases}
\end{cases}
$$

Two types of metal binding are in common use, viz.: flat straps and wires.

Flat straps are either annealed or unannealed. The differences in the properties of these two kinds of metal strapping result from a special heat treatment given unannealed strapping, thus producing annealed strapping.

Annealed strapping, as a result of this treatment, has a much lower tensile strength, stretches considerably more before failure or under a given load than unannealed strapping of the same size and is also more easily penetrated by nails. There are many special types of annealed strapping such as the plain, embossed, and corrugated, as well as various other types, that have holes or slots cut to receive the nails.

Wire ordinarily used for box construction is annealed. It is used either as a single strand, ordinarily held in place by twisting the ends, or in two or more strands twisted together and held in place by nails driven through spaces between the wires.

Plain unannealed strapping is generally applied by fastening the overlapping ends with a seal. The seal should provide a joint whose strength is nearly equal to that of the strap.

Straps nailed around the extreme ends act somewhat as a cleat in resisting skewing or weaving of the box; retard the nails pulling from the ends; prevent the nails in the straps from pulling the heads through the sides, top, and bottom; and assist in preventing the nails from shearing out at the ends of the boards by acting in the nature of washers under the nail heads. This method of reinforcement also gives more secure nailing because the nails driven through the straps are in addition to those ordinarily used in the manufacture of the box. The strapping should be held in place with nails of the same size as those used to hold the sides, top, and bottom to the ends.

Nailless straps placed some distance from the ends of a box absorb considerable shock which is ordinarily transmitted to the sides, top, and bottom, and thus relieve the direct pull on the nails in the end and also reduce failures due to the sides, top, and bottom splitting or breaking across the grain. Nailless straps do not add as much rigidity to a box as nailed straps and have less value in reducing shear on the nails in the ends of the sides, top, and bottom. In some cases nailless straps permit the use of thinner material in the sides, top, and bottom than is permitted by straps nailed around the end of the box.

Either annealed or unannealed strapping may be used when nailed around the box, whereas, only unannealed strapping should be used when held in place with a seal.

Metal bindings, particularly the nailless variety, to be most effective, should be drawn sufficiently tight to cut into the corners of the box, and maintained under considerable tension until the box has served the desired purpose.

One method of retaining the tension in nailless strapping is by building the box in such a manner that neither the top nor bottom laps the sides. The tension of the strapping when drawn snug is sufficient to spring the sides, top and bottom of the box in against the contents so that the edge boards overlap near the center. As a result the middle of the box is smaller than the ends, and the straps will not slip off, even though the box shrinks.

Another method of applying straps between the ends is to put battens[1] on each face as shown in figure 1, Plate V, and then nail the straps to the battens. The battens make it possible to use longer nails without injuring the contents. In some instances, the straps may, in the absence of such battens, be nailed directly to faces of the box, if the material packed will not be injured by the nail points.

Shipping containers are frequently subjected to adverse moisture conditions and for that reason the metal bindings should be treated to resist rust.

Boxes having comparatively thin sides, top, and bottom

---

[1]See page 62 for discussion of battens.

and bound with metal bindings often are more serviceable than those constructed of heavier lumber without such bindings. In some cases it is possible to reduce the thicknesses of material 20 per cent or more and at the same time permit the use of a poorer grade of lumber when metal bindings are properly used, without any reduction in serviceability of the container.

REINFORCEMENTS AND HANDLES

Corner Irons, Hinges, and Locks—Corner irons may be necessary on boxes for certain conditions of service, such as returnable boxes and heavy chests. The style of iron to be used will depend entirely on the conditions of service.

The important requirements of hinges and locks are that they hold the cover securely in place, do not project so as to interfere with handling and stacking, and do not easily get out of usable condition. Locks ordinarily should be of such design that they may be held in a closed position by a seal rather than a key. Hinges should allow the cover to open until the free edge lies in a plane parallel to the bottom of the box, so as to minimize the danger of breakage, or injury to the box.

Battens—Battens and cleats are very much alike. Either or both are put on some styles of boxes when they are made, or later to give additional strength. They may be on the inner or outer surface of a box. (See Plates III and VIII.) Cleats and battens on the outer surface usually increase the displacement and increase the cost, and often interfere with the stacking of the boxes. They should be used, therefore, only when no other method of construction can be devised which will be as economical and as satisfactory.

Hand-holds and Handles—It is desirable to have hand-holds or handles on many boxes, such as those to contain heavy commodities which must be handled with reasonable care, and commodities of delicate construction like scientific instruments. Returnable boxes should also have hand-holds or handles, as this construction tends to encourage careful handling.

Two common types of hand-holds are shown in figures 1 and 3, Plate VI. Being near the points of the nails holding the tops to the ends they weaken the ends in such a manner as to increase the danger of failure along the lines passing through the length of the hand-holds.

One method of arranging a hand-hold on boxes of Styles 2, 2½, and 3, Plate III, is to shape the lower edge of cleats 5-1 and 6-1[1] as shown by the section in figure 2, Plate VI.

A method of applying rope handles to boxes is shown in figure 4, Plate VI. Three methods of fastening the cleats to the ends have been used as follows: four No. 11 flat-head wood screws, 1¾ inches long, six 7d cement-coated nails driven through and clinched, and four No. 11 drive screws, 1¾ inches in length. Wood screws must be properly driven with a screw driver[2], as driving with a hammer to any appreciable depth seriously reduces their holding power. The cleats are 1⅛ inches thick and the ends $\frac{13}{16}$ inch thick. The rope is three strand and is ⅝ of an inch in diameter. Both types of screws are arranged as shown in figure 4, Plate VI. It should be noted that in each case two nails or screws pass through that part of the rope within the vertical groove of the cleats and the lowest of these two screws or nails is about 2 inches from the end of the rope. The grooves in the cleats should be of such size that the rope is squeezed when the screws or nails are driven home. Tests consisting of a pull on the rope in a direction perpendicular to the box ends have shown that the nailed construction (six 7d cement-coated nails per cleat) offers the greatest, and the drive screw (four No. 11 drive screws 1¾ inches long) the least resistance to failure, the wood screws (four No. 11 flat-head wood screws 1¾ inches long) giving intermediate results.

Webbing has been recommended as a substitute for rope in handles. One style is shown in figure 6, Plate VI. The end piece of the box is slotted and the webbing is passed through these slots and nailed on the inner surface of the box. Several large-headed nails should be used and may be driven through and clinched. Handles of this type have sustained an average direct pull before failure of 800 pounds in a direction perpendicular to the end of the box.

Another method of attaching webbing handles is shown in figure 5, Plate VI, in which the webbing is passed through ½-inch round holes and the ends secured by nails. Such handles have sustained a direct pull of 700 pounds perpendicular to the end of a box. These handles are made of seven-cord cotton rein webbing ⅛ inch thick and 1¼ inches wide. One of their advantages over rope handles is that cleats are

---

[1]See figures 3 and 4, Plate XIV.
[2]See table on page 59.

not needed to hold the webbing and, therefore, it is possible to construct a box with greater volumetric efficiency.

Many styles of metal handles can be obtained from manufacturers of trunk and chest hardware. One objection to many types of metal handles is that in case a box or chest drops on an end, the handle usually extends in such a position as to get the full force of the impact, which in many instances crushes in the end of the box or breaks the handle.

CHARACTERISTICS OF THE VARIOUS STYLES OF BOXES

Nailed Boxes—In Style 1, Plate III, boxes, the grain of the ends and sides runs approximately parallel to the top and bottom surfaces. One of the common failures in this type of box is splitting of the ends and sides, or failure of the joints in these parts, since the only resistance to such failures lies in the strength of joints,[1] if present, or the strength of wood in tension across the grain, which is not large and is extremely variable in any species of wood. The smaller holding power of nails driven into the end grain of wood, i. e., with their shanks parallel with the grain, as compared to side grain nailing is a source of weakness in the joints between sides and ends.

To improve on Style 1 ends and guard against the liability of complete failure from splitting of ends and sides, rectangular and sometimes triangular corner cleats (Style 5) are added inside when the character of the contents permits. This construction does not increase the displacement of the box and is, therefore, not objectionable in that respect. If these cleats can be made large enough the sides may also be nailed to them, which of course increases the strength of the nailing at this point. These inside cleats should be shorter than the inside depth of the box, so that if the sides and ends shrink[2] the cleats will not cause an opening of the joints between the bottom and ends

The most common method of preventing box ends from splitting and of supplementing the holding power of nails driven in the end grain is the addition of two outside cleats on each end as shown in Style 4, Plate III. These cleats should be long enough to come nearly flush with the outer surfaces of the top and bottom. They will thus aid in keeping the top and bottom in place and will also take some of

---

[1]For discussion of joints see page 43.
[2]See page 22.

the thrust if a box is dropped on a corner. If a box is constructed with the ends of these cleats made exactly flush with the outer surfaces of the top and bottom, and shrinkage occurs later, it may cause the ends of the cleats to project beyond the top and bottom; and they may be pulled loose if the box is slid in such a way that the ends of the cleats catch on some object. This failure is more apt to happen with heavy boxes and with boxes that are handled in chutes and slides. The amount that the cleats should be cut short to allow for shrinkage depends on the moisture content of the lumber when the box is constructed and the storage conditions afterward[1]. Usually an allowance of from $\frac{1}{8}$ to $\frac{3}{16}$ of an inch at each end will be sufficient.

The reason for the two additional horizontal cleats 5-1 and 5-3[2] on Style 2 (Plate III) is to give increased nail-holding power to the end of the box. The greatest increase in nail-holding power will be obtained when the cleats are made of the denser woods. The usual failure in these ends is a split along the inner edge of cleats 5-1 or 5-3, which allows a cleat with part of the end board to pull away with the top or bottom. The resistance to such failure is due to the strength of the end board in tension supplemented by the action of the vertical cleats. In Styles 2 and 2½ it is possible to get more nails in the vertical cleats near the top and bottom edges of a box than in Style 3, which more effectually prevents the box end from splitting adjacent to the edge of the horizontal cleats.

The vertical cleats in Styles 2 and 2½ should also be cut slightly short, so that if shrinkage occurs in the end pieces these cleats will not hold the top and bottom and allow the end boards to pull away. Style 2½ also has the advantage that when the bottom and top are being nailed to the ends the notches or steps on the vertical cleats take the thrust that would otherwise come on the nails holding the horizontal cleats. In some cases, as when driving several nails into a cleat made of dense wood, this thrust is very severe.

In manufacturing boxes with square ends Style 3 has the advantage that all four cleats are the same length, hence interchangeable. When a very symmetrical end is desired rather than the strongest end the mitred cleats are preferred.

Inasmuch as one of the chief functions of cleats in many

[1]See page 22.
[2]See figures 3 and 4, Plate XIV.

boxes is to provide additional nail-holding power in the box ends, it is desirable that the cleats and end boards be the same thickness, so that the same sized nails may be used in them.

The Hardware Type of "three-way" corner construction of box shown in figure 3, Plate XIV, is largely used for hardware. This type is well adapted to boxes carrying heavy loads; and for boxes in which the maximum dimension is not more than two times the minimum dimension. It is better practice, however, to have the dimensions as nearly equal as possible. When the difference between the minimum and maximum dimensions becomes excessive some other style of box or crate should be used. All the faces must be made of material of the same thickness; and since all nails are driven into the edges of boards constituting the faces of the box, it is necessary that the material be thick enough to prevent its being split by these nails. It is not considered feasible to make a box of this type out of material less than ½ inch in thickness.

An advantage of the "three-way" construction is that each face has nails driven in it in two directions, those driven through its ends perpendicular to its face being at right angles to those driven into and perpendicular to its edges. A further and probably greater advantage is that nails are driven into the side grain of the wood.

One objection to the "hardware type" of box is that in closing it after packing, four edges must be nailed. The boards meeting at these edges are so arranged that nails must be driven in three directions, which makes much turning and handling of the box necessary, especially if a nailing machine is used for the work. In closing other types of nailed boxes, the nails are all driven in one direction even though they may be distributed along four edges of one face.

Lock-Corner Boxes—The box shown in Style 6, Plate III, is of a type in which the sides and ends are joined by a series of tenons which interlock and are called "locks."

These locks are held together by gluing and the top and bottom are fastened in some other way. This method of construction allows the use of thinner material in the ends than is possible with a nailed box (Style 1) properly designed for carrying the same contents. The lock corner, if properly glued, gives a more rigid box than nailed corners, there being no appreciable distortion before failure occurs. In lock-corner boxes there is danger that the ends and sides will split open, or that failure will occur in the joints which may be present in these ends and sides.[1] Since the ends can be made of thinner material in the lock-corner boxes than in nailed boxes, there is danger that the desire to save material may be carried to extremes, with the result of making the end too thin for properly nailing on the top and bottom.

Tests show that in lock-corner boxes a considerable number of the failures occur because the nails pull from or split the edges of the thin ends, locks open, ends split, and matched joints lack sufficient strength. The ends of lock-corner boxes are longer than those of nailed boxes because of the additional length necessary for forming the locks. This extra length is equal to twice the thickness of the side boards plus enough for trimming after gluing. The sides also need enough extra material for trimming after gluing.

Dovetail Boxes—The dovetail construction shown in figure 2, Plate VII, is used to a limited extent for expensive boxes, returnable boxes, chests, etc., but such construction is more common in cabinet work and furniture. The top and bottom of the boxes may be fastened by any desirable method. In figure 1, Plate VII, are shown the tenons as they appear on two pieces before being joined to form a corner.

One point of superiority of the dovetailed corner in comparison with the lock-corner is the advantage of the wedging action in preventing failure as described in connection with Style 6, Plate III, but it requires more complicated machinery to produce the dovetailed corner.

The disadvantages due to thin ends, joint failures, etc., in dovetailed corners are very similar to those in lock-corners.

Wirebound Boxes—The stock in the sides, top, bottom, and ends of the boxes shown in figures 1 and 3, Plate VIII, is called sheet material and it is usually made of rotary-cut stock, although resawed lumber is used occasionally.

Figure 4, Plate VIII, shows a mat for a "4-one" box as delivered by a stitching or fabricating machine ready to be assembled with the end panels, or shipped in this form to the consumer. The cleats are held to the sheet material by the staples which pass over the wires, through the sheet material, and have their points firmly held in the cleats. Staples not driven into cleats are clinched on the inside surface of the sheet material. Staples over all wires should be spaced from

[1]See page 43 for discussion of matched joints.

1¾ to 2¾ inches apart. It has been found that with 1¾-inch spacing, the end binding wires do not slip off the corners of the box as frequently as when the spacing is larger. The end pieces are stapled or nailed on the inside surface of three of the cleats when the box is assembled, the remaining cleat on each end being attached only to the top as shown in figure 6, Plate VIII. The binding wires are twisted together near one edge of a side, to close the box.

There are several styles of end joints for the cleats in wirebound boxes. The mortise and tenon now in general use and the step-mitre are shown in figure 4, Plate VII. The step-mitre is the older method but has at the present time been almost entirely abandoned in favor of the mortise and tenon joint. Cleats with plain mitred ends are also coming into use. They are more economical to manufacture than other types and are very satisfactory for containers for some commodities.

An advantage of the step-mitred and plain mitred cleats is that they allow staples to be driven much nearer the ends of the cleats, which aids materially in preventing the binding wires from being forced off the corners of the box. An advantage of the mortise and tenon joint is' that it prevents lateral movement of the cleats respecting each other and thus makes a box which is more rigid than one made with either style of mitred cleats.

Cleats for the smaller wirebound boxes are usually made approximately ¾ by $1\frac{3}{16}$ inch in cross section. The resistance of the box to destructive hazards encountered in service is only slightly improved by increasing the width of the cleats to $1\frac{3}{16}$ inches.

The end construction of wirebound boxes may be materially strengthened by the addition of battens. The size, number, position and method of fastening battens depends very largely on the severity and location of the thrusts transmitted to the ends by the contents of the boxes. An arrangement of battens which has been tested at the Forest Products Laboratory is shown in figures 1 and 2, Plate VIII. These battens protect and strengthen the end material, increase the rigidity of the box, add strength to the box to resist vertical compression, such as occurs when a box is subjected to an exterior load as in the bottom layer, make the corner construction more rigid, and tend to maintain the relative position of the cleats in the step-mitre construction, and also in the mortise and tenon construction after either the tenon or the sides of the mortise have failed. The amount of strength added by these battens increases as the width of the battens increases to a maximum, varying from 1½ to 2 inches. The thickness remains constantly equal to that of the cleats. Battens much smaller can not be properly nailed and neither larger battens nor solid ends add appreciably to the serviceability of a box,

Fig. 21—Nailed box showing panel style of construction.

as the ends are then too strong and out of balance with the other parts. In any case, battens must be securely nailed in order to get the greatest increase in strength. The battens on the Fassnacht type of box, figure 1, Plate VIII, have a tenon on each end and a tongue on one edge which fit into grooves in the edges of the cleats. This construction materially assists the nails in holding the battens in place.

Inside liners or corner cleats, as shown in figure 5, Plate VIII, strengthen the box to some extent.

The sheet material of wirebound boxes is rather easily punctured by the corners of other boxes and projecting objects, and is not adaptable for commodities which are liable to be seriously injured by such hazards.

Wirebound boxes are made of thin material, and are, therefore, economical in the use of wood. They can be shipped in the mat (figure 4, Plate VIII) or knocked down form, and can be assembled at the point of filling with less work than is required to nail shooks[1] together. They weigh less than other wooden boxes adaptable to the service and made of the same wood. Special tools have been devised for efficiently twisting the wires for closing wirebound boxes.

Panel Boxes—Boxes of the type shown in figure 21 are used to a considerable extent. The panels are made of plywood, three-ply stock being most generally used. The plywood is nailed to cleats or battens at each edge to form the panels and these panels are then nailed together through the cleats to form the box. This construction makes a box of light weight which if properly nailed is much tighter and more rigid than a wirebound box. Since the panels are of plywood, the weakening effect of a defect which occurs in one of the plies is largely annulled by the adjacent sound material of the other plies. The construction is, therefore, more uniform than any type of box made of the common grades of box lumber or single-ply veneer.

Plywood equal in thickness to lumber offers much more resistance to puncturing and therefore is more desirable for boxes carrying commodities which will be damaged by that hazard. Plywood also shrinks very little, and does not split or pull away from nails.

FACTORS DETERMINING THE AMOUNT OF STRENGTH REQUIRED

Contents—With contents made up of an equal number of units, identical in size and shape, having similar properties but different weights, and packed in boxes of the same inside dimensions, it is evident that the heavier commodity will require a stronger box to withstand equivalent amounts of rough handling.

The details of a box to give this additional strength can be most readily determined by a series of tests on various boxes of different design. Contents which will themselves absorb considerable shock will materially prolong the life of

---

[1]The ends, tops, bottoms and sides of wooden boxes before assembling are called shooks.

a container. It has been demonstrated by tests that a box containing 60 pounds in No. 3 food cans will fail more quickly under test than an identical box filled with the same weight of sand and sawdust in bulk.

Thus the nature of the commodity or inner containers has a marked influence on the degree of strength required in a box for performing a specified service.

If a box is packed with contents consisting of a few units of rectangular section there is a tendency for these units to maintain their relative positions and thus prevent bending of the box boards because of the arching effect produced.

Boxes for carrying one rigid object of rectangular shape need form only a protecting envelope with perhaps little strength. Thus the strength of a box in some respects may be diminished in proportion to the amount of resistance offered by the contents to injury and deformation.

Hazards of Transportation—Nothing has more influence on determining the degree of strength required in a box than the hazards of transportation. They usually tend to cause failure in boxes by one or more of three actions, viz., weaving or wrenching, puncturing or breaking various parts, or collapsing.[1] The collapsing action may occur as diagonal compression between opposite corners or opposite edges, or as compression perpendicular to the ends, sides, top, and bottom. Boxes which are dropped, thrown, and rolled when being handled by hand may encounter all of these hazards, the severity of which will depend on the care exercised in doing the work. The hazards that boxes are subjected to in being conveyed by motor trucks in long distance transportation may result in considerable twisting, weaving, and jamming of the boxes, and there may be destructive compressive stresses transmitted to the boxes at the bottom of the load.

The hazards of shipping by freight are at times very severe, especially those occurring during the switching and making up of trains. In cars containing a miscellaneous lot of commodities, loaded with little thought of proper arrangement and blocking so that the stronger packages should receive the severest strains, there will most surely be a large loss from damaged goods. The contents of cars, if not well braced in position, receive severe weaving and wrenching strains and may also be subjected to serious compressive and puncturing stresses.

---

[1]See page 60.

Destructive conditions due to the elements, especially moisture, are apt to be experienced by boxes in transit. Such conditions may be encountered at loading points where platforms and wharves are not covered, at prepaid shipping points in the country where there are no railroad agents, and in transit by boats, barges, trucks, wagons, pack animals, and refrigerator cars.

The strength requirements for export shipment are much greater than for domestic shipment; in fact, export shipments may undergo all the hazards of domestic shipments previous to arriving at the wharf for loading; and after reaching a foreign port there may be a long journey inland.

The stevedores who handle export shipments pay little attention to proper handling of freight, with the result that weak packages and those containing fragile materials are tossed and thrown about with heavier and stronger ones. Cargo hooks are indiscriminately used on packages, which are punctured by them and the contents injured. In many foreign ports the stevedores are people who can not read directions that may be printed on packages regarding the nature of the contents and directions for handling, and thus all containers are treated much alike.

One method of loading and unloading ships is by the use of cargo nets. These are large nets on which a quantity of boxes are piled and the corners of the net then drawn together for lifting in loading and unloading. In these operations the boxes are thrown violently together in the nets and in swinging into place over the ship or lighter the load net often strikes severely against the side of the ship, or some other object, with resulting injuries to the contents of the net. In emptying the net, one edge is often released several feet above the deck, the contents falling the remainder of the distance, accompanied by a rolling action. Boxes must be well constructed to endure such service.

Rope, slings, chains, and grappling irons are used on large, heavy boxes. The chief hazards connected with their use are dropping due to premature releasing of the hoisting device, crushing and bending action due to the manner of applying slings, etc., and striking against other objects in swinging from one position to another. Such hazards are very common and provision must be made for resisting them in boxes for export shipment.

In many harbors the ships must be unloaded by means

of lighters. This means extremely severe handling, especially at ports where a rough sea is common. An extra handling is then necessary from the lighters to wharves.

The hazards of a sea voyage are the stresses resulting from improper stowage in the ships, the shifting of a cargo, and weakening effects due to change in moisture content and corrosion of metal parts.

FACTORS DETERMINING THE SIZE OF A BOX

Gross Weight—The gross weight for boxes should, if possible, be such as will make a reasonable load for one man or, in some instances, for one woman. The French Government in connection with war work fixed the maximum load for a woman at seventy pounds. When the load must be more than one person can handle efficiently it should, if possible, be increased to make a reasonable load for two. If the gross weight is too much for one person and too light for two the work of handling can not be so efficiently done. The weight of the container should be as small as proper design will permit with the object of saving in freight charges and box material.

Desired Quantity—With some commodities it is the quantity which is ordinarily desired by the consumer which will determine the size of the box to be used. With small boxes, however, several of them may be in turn packed in a larger box or crate to facilitate handling and to give better protection in shipping.

Nesting, Disassembling, or Knocking Down Contents—For some objects, the size of the box must be larger than is ordinarily desired for convenient handling; in such cases, as much nesting and disassembling of parts should be done as is deemed advisable in order to reduce the size of the box. Disassembling will often allow parts that would be easily broken to be packed more securely.

Minimum Displacement—In every case, the amount of space required for a box, i. e., its displacement, should be as small as possible. This is especially desirable for export shipments, as rates are based on the space required rather than the weight. Probably the space requirement will enter more directly into rates for domestic shipments in the future. Minimum displacement also means in most instances the use of a smaller amount of material in constructing the boxes and a reduction in storage space required for them, both in shook and assembled form.

Traffic Limitations—There are certain traffic limitations and regulations which influence the design of some boxes. The design of boxes for shipping explosives and dangerous articles by freight is regulated by the Interstate Commerce Commission. The size and weight of packages for shipment by parcel post is limited. More definite regulations by traffic officials regarding the method of boxing and packing certain commodities would help to solve the problem of obtaining more adequate shipping containers.

SPECIAL CONSTRUCTIONS

Protection of Fragile and Delicate Contents—Many commodities are extremely susceptible to damage in transit; too much information can not be had by box designers regarding methods of packing such commodities correctly.

Protection against injury by moisture must be provided for some commodities. One method is to provide a water-resisting paper liner, or a tight sheet-metal liner. When these liners are to be used, the space which they require must be provided for in the design of the box.

Some kinds of merchandise need to be packed in materials that will not transmit to the contents the shocks and impacts received by the box. Among such packing materials are corrugated fiber and straw board, excelsior, straw, hay, shavings, sawdust, etc. The amount of space required for these cushioning materials will depend on the fragility and weight of the commodity and the severity of the hazards of transportation.

One method of preventing serious stresses and shocks due to boxes falling on their corners is to provide each corner with a bumper of cushioning material. Another method is that of "flotation," which consists of packing one box within another, with the intervening space between all faces of the boxes filled with a cushioning material or some system of mechanical spring supporters. In this problem two boxes of proper relative size and strength must be designed and the cushioning feature must be provided for.

The various individual elements constituting the contents of some boxes need to be prevented by separators of some character from jolting against each other. One method is to form a series of cells or individual compartments for each element of the contents, as shown in figure 22. Tests have demonstrated that for carrying hand grenades cells made from corrugated fiber board sustaining 175 pounds per square inch before puncturing, as recorded by the Mullen, the Webb and other similar testers, are more efficient than wooden cells

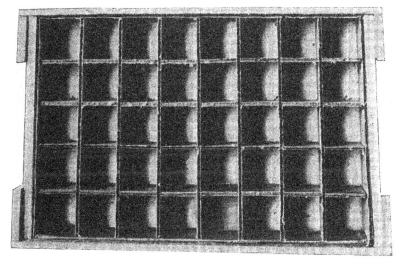

FIG. 22—Box with corrugated fiber board lining and cells.

made from $\frac{5}{16}$-inch lumber. The wooden cells broke up sooner and absorbed less shock than the corrugated fiber board cells and thus they transmitted the shocks of the contents to the container, thereby stimulating the action tending to injure the contents.

The character of the material composing the cell will depend on the nature of the contents to be packed. Some contents need to be protected from surface abrasions only, and in such cases it is the cushioning property that is desired in the cell walls; for other contents, much strength may be needed in the cell walls, as the contents may not be strong enough to withstand the pressures which occur within the container.

Whatever the material of which the cell is made its dimensions should be such as to permit the contents to fit snugly; this diminishes the force of the impacts tending to destroy both the cells and the container, and also allows the total displacement of the container to be reduced.

In designing boxes for some commodities the more fragile and delicate parts must be separated from the stronger parts by partitions of considerable strength. In some instances, the amount of material in a box can be reduced and a more satisfactory container secured by using one or more partitions. Partitions may be placed horizontally as well as vertically and also may be left unfastened so as to facilitate removal. Some articles can be shipped to advantage by using trays for supporting them in their boxes. Trays may be constructed of plywood or of lumber. In using lumber for trays in some instances it is well to put splines in their ends to prevent splitting and, to some extent, warping. There is much less change in size and shape of symmetrically constructed ply wood than lumber when subjected to change in moisture content. It may be desirable at times to use sheet metal, wall board, etc., for trays.

When several articles of varied shape with some delicate parts attached are to be packed in the same box, a series of internal braces, supporters, separators, etc., will have to be designed. Good examples of such boxes are those for carrying rifles, machine guns, valuable tools, scientific instruments, and machines.

Vermin—Some commodities are attacked by various species of vermin in storage and transit, especially during sea voyages. It may devolve on the box designer to devise a style of box construction or inside lining which will resist the ravages of such destructive pests.

Thieving—Great losses occur from theft while goods are in transit. The problem is a very serious one and much attention should be given to the design of boxes which can not be readily opened and reclosed without detection.

The prevention of thievery is largely a matter of police protection. Seals, straps, and more rigid construction as described in other parts of this book are valuable in that theft is more quickly discerned, and the thief is deterred by his knowledge of the box construction.

CHAPTER III

CRATE DESIGN

FACTORS AFFECTING STRENGTH OF CRATES—FACTORS DETERMINING AMOUNT OF STRENGTH REQUIRED IN CRATES—FACTORS INFLUENCING THE SIZE OF CRATES.

Many of the factors influencing the details of design for boxes will similarly affect the design of crates. Among these are the availability, supply, and cost of lumber, manufacturing limitations, and balanced construction which are discussed in Chapter II.

FACTORS AFFECTING STRENGTH OF CRATES

INFLUENCE OF STYLES OF CRATES ON STRENGTH

The serviceability of crates is vitally affected by the style of construction, especially the method of joining the frame members at the corners of the crates and the kind of fastenings used.

Types of Corner Construction—In Figures 1, 2 and 3, Plate XI, are shown three types of the "three-way" corner construction for joining the frame members of crates. If only the types of construction are considered, that shown in figure 1, Plate XI, is the most desirable because all of the frame members are fastened in the same way. In figures 2 and 3, Plate XI, the arrangement of the members is different; but these types may be desirable when crating some objects. Material which is approximately square in cross section is preferred in these two constructions. There are sixteen different variations of the three-way corner. (See Plate XII.)

An advantage of the three-way corner is its symmetrical construction in which the members are fastened together in such a way that through each member nails or bolts may be passed in two directions at right angles to each other, thus uniting the members securely and reducing the danger of their being split by the bolts or nails; nails or bolts passing through the members in one direction resist the splitting action of those at right angles to them.

Another advantage of the three-way corner is the arrangement of the members so that only one thickness of frame material intervenes between the contents and the outer surface of the crate, thereby in most cases keeping the displacement lower than is possible with other styles of crates

In figure 5, Plate XI, is shown a type of crate-corner construction which is largely used but which is inferior to the three-way corner construction in at least two respects; viz., there are two thicknesses of the material outside the object on two faces of the crate, which increases the total displacement of the crate, and some of the members have nails through them in only one direction, so that they are not held so securely to the other members as they would be with the three-way corner construction. The addition of a second piece along one edge, as shown in figure 6, Plate XI, gives additional strength to the corner, makes the combined member stiffer and stronger, and, in some instances, gives additional support and protection to the contents.

Frame Members—The frame members of a crate as shown in figures 3 and 4, Plate XIII, constitute the foundation to which all other parts are connected either directly or indirectly. The frame members must have sufficient size and strength to form a foundation skeleton upon which to complete a crate that will carry the contents to its destination with little danger of injury, even though severe hazards are encountered. Not only the vertical and horizontal members should be strong enough to support the exterior loads to be put upon a crate in storage or transit, but also the diagonal and cross bracing[1] must be sufficient and so arranged as to distribute the stresses and hold the other crate members in proper position. Without proper bracing it is practically impossible to build a crate that will not weave or skew in transportation, even though the three-way corner construction is used and the crate when freshly made appears quite rigid. This skewing and weaving is largely responsible for such damage as rubbing of varnished surfaces, and the breaking of legs and other projecting parts of furniture.

The amount that any piece will support in compression, considering that it is a column so short or so firmly braced that there is no danger of failure by buckling or bending, will

---

[1]See page 84 relative to internal braces.

be found by multiplying the area of the cross section of the piece in square inches by the safe allowable load for the material in pounds per square inch.[1] The safe allowable load is always considerably less than the ultimate load that a material can support. The horizontal top members must be strong enough to sustain between points of support such bending loads as may be placed on them in storage or shipment.

Skids—The lower horizontal frame members running lengthwise of a crate usually form the skids. The skids support the contents directly, or indirectly through intervening members, and also the weight of the crate and superimposed loads, unless the lower ends of the vertical members rest on some other support. For heavy objects, which must be moved on rollers and hoisted with chains or slings, the skids should have extra pieces added to them as shown in figure 4, Plate XIII, which will increase the bending resistance and provide a bearing surface for the rollers, chains, etc. The ends of these additional pieces should be beveled or scarfed to facilitate sliding or the passage of the skids onto rollers, and should also extend outward underneath the vertical members of the crate to support them as indicated, for otherwise any load put on the crate will produce compression and shear on the fastenings which connect the vertical members to the skids. These extra pieces must also be securely fastened throughout their length to the frame members in order to make the combined parts act more nearly as a single piece skid, which increases the resistance to horizontal shear and bending. When a crate is being moved, it may at some time be supported by rollers or slings in such a way as to produce serious bending moments in the skids. To calculate the weight that a skid will support when resting on a roller midway between points of loading, the formula for beam loading given in U. S. Department of Agriculture Bulletin 556, "Mechanical Properties of Woods Grown in the United States," may be used.

Bracing Long Crates—When, in a long crate, each skid and the top horizontal frame member are securely fastened together by several cross members and substantial diagonal or cross braces are provided, a truss-like form of construction is obtained which greatly increases the resistance of the crate to bending. A side view of such a crate is shown in figure 1,

---

[1] See United States Department of Agriculture Bulletin 556, "Mechanical Properties of Woods Grown in the United States," 10 cents per copy, obtainable from Superintendent of Documents, Washington, D. C.

Plate XIII. To obtain the number of cross members and number of angular braces or sets of cross bracing for any face of a crate, use the following rule:

Divide the longer dimensions of any face by its shorter dimension, and,

1. If the result is less than one and one-half use one angular brace or one set of cross bracing;

2. If the result is one and one-half or more and less than three, use one cross member and two angular braces or two sets of cross bracing;

3. If the result is three or greater, use a number of angular braces or sets of cross bracing equal to the largest whole number in the result. The number of cross members will be one less. (See figures 1 and 2, Plate XIII.) This method of figuring the number of braces to be used divides the face of the crate into panels which are approximately square, the braces making angles of about 45 degrees with horizontal members, which is considered the best practice.

Fitting and Fastening Braces—In cutting the end of a diagonal or cross brace, the toe should be made with a flat surface or butt against the adjacent member. (See figure 4, Plate XI.) Care should be exercised, however, that the distance from the toe to the heel is great enough to provide for properly nailing or bolting the brace.

The thickness of a diagonal brace, or a set of cross braces, figure 4, Plate XIII, plus the thickness of sheathing when used, figure 2, Plate XIII, should not exceed the thickness of the frame members.

When cross bracing is put on as in figure 4, Plate XI, the outside brace should have blocks between its ends and the frame members to which it is nailed unless the length of the brace is such that the amount of bending produced by omitting the blocks will not seriously strain the braces. The initial bending produced by omitting the blocks will tend to increase the danger of failure by buckling when a brace is subjected to endwise compression. Such blocks as are used under braces should preferably extend for some distance along their length and be securely fastened to them to minimize the danger of the blocks being split and getting out of place. Cross braces should be securely fastened together at the point of intersection. Nails driven through and clinched, or bolts, are preferable for this fastening.

Scabbing—In figure 2, Plate XIII, the use of scabbing is shown. Scabbing consists of a piece nailed across a joint on the faces of two pieces to unite them securely. It is similar to a plaster joint in a timber construction. The chief requirements of pieces which are to be used for scabbing is that they possess sufficient strength and resistance to being split or sheared by nails or bolts. It is well to have the scabbing in this particular construction wide enough to support the sides of the toes of the braces, as indicated in the figure.

Sheathing—The purpose of sheathing is to protect the contents from the elements, reduce losses of small parts by thieving, and prevent injuries to contents from external objects. Sheathing when securely nailed also strengthens a crate to a considerable extent, especially if it is run in a diagonal direction. It may be put entirely outside of the frame members and bracing. A poorer grade of material can be used for sheathing than for the other parts of a crate. Usually matched lumber surfaced at least on one side and with the surfaced side exposed to receive shipping directions and advertising is preferred.

Battens on crates are used for additional supports for sheathing and for holding on waterproof coverings.

PHYSICAL PROPERTIES OF WOOD AFFECTING STRENGTH.

Relative Thickness of Material—For general crating work the harder woods of Groups 2, 3, and 4[1] may be 25 per cent less in thickness than material from Group 1 to give approximately equal strength.

Moisture Content—The moisture content in crate lumber should be within limits of from 12 to 18 per cent. Decreases in moisture content after construction will loosen the fastenings and joints, cause members to check, allow internal braces to become loose, and diminish the effectiveness of the sheathing as a protection to the contents.

Defects—In crate members and braces, defects must be more rigidly excluded than in lumber for boxes, as the parts must have more uniform strength. Except in special cases, such defects as are allowed in box material can be permitted in crate sheathing. Defects are discussed fully in Chapter II.

Nailing and Bolting Qualities of Wood—Since the main fastenings in crates come at the ends of the various members it is important that lumber which is not easily split by nails

---

[1]See page 100 for grouping of woods for box construction.

or bolts be used. It is a well-known fact that in many wooden structures the great weakness and danger of failure is due to inability to get the ends of the various members in tension fastened in such a way as to stress them even to a point of safe loading. For crate construction, lumber which is warped or twisted is more objectionable than similarly affected material would be in box work. Considerable initial stress is produced in a seriously warped timber when it it is forced to assume approximately a normal shape. This necessarily reduces the ability of the piece to resist external forces.

The various factors which influence the holding power of nails and the strength of nailed joints are discussed on pages 51 and 101.

Bolts, when holes of proper size are bored for them, do not produce a wedging action which tends to split the members of a crate as do large nails or spikes. The clamping action produced by bolts holds the members together more securely than nails, which are dependent in such action upon the friction of their shanks in the wood of the member holding the points. In case of shrinkage and splitting of the wood, bolts may be tightened and will continue to be more effective than nails under similar conditions.

Fastenings and Reinforcements—Because of the open construction of crates the total amount of space allowed for fastening is less than for boxes of the same size and therefore the fastenings on crates must be relatively stronger. Much of the discussion, however, pertaining to nails in the section on "Fastenings and Reinforcements," page 53, will apply to the nails used in crating work.

Nails and Nailing—The size of cement-coated nails recommended for the various thicknesses of lumber used in all parts of a crate is given in Table 13.

Frame members and braces should have not less than two nails in any end and as many additional nails as can be driven without weakening the joints by splitting the members. Nails in sheathing should be staggered, the distance between their centers, measured along the length of the piece nailed to, being equal to $\frac{1}{4}$ inch for each penny of the nails used. Sheathing should be nailed to all members of the crate which it crosses.

It is often supposed that driving nails at a slant results in an increased resistance to withdrawal and to shear in the joints. While in some special cases this belief is supported by test results, tests show that in most cases slanting causes a loss of efficiency. The efficiency of nails and the number of nails that can be used without splitting can be very considerably increased by boring holes. The boring of holes gives definite bearing

TABLE 13. SIZE OF NAILS FOR CRATING

Thickness of lumber in inches		Penny of cement-coated box nails[1]	
Against nail head	Holding nail point	Group I woods[2]	Group II woods[2]
1/2	1/2 to 5/8	6	5
1/2 to 5/8	3/4 and over	7	6
3/4	3/4 and over	8	7
13/16 to 7/8	13/16 and over	9	8
1	1 and over	10	9
1 1/8 to 1 1/4	1 1/8 and over	12	10
1 3/8 to 1 1/2	1 3/8 and over	16	12
1 5/8 to 1 3/4	1 5/8 and over	20	16
1 7/8 to 2	1 7/8 and over	30	20
2 1/8 to 2 1/4	2 1/8 and over	40	30

on the end grain of the wood, whereas driving without holes forces the wood fibers aside without affording such definite bearing. Nails driven in holes slightly smaller than their diameter have considerably better resistance both to direct pull and to shear in the joint than nails driven without holes.

Under the repeated shocks and more or less constant weaving action to which crates are subjected slender nails bend near the surface of the pieces joined, and without loosening the friction grip toward the point of the nail. As the diameter of the nail is increased the stiffness also increases, and at a much more rapid rate, and the deformation of the wood and decrease of the friction grip progresses toward the point of the nail. Consequently, the larger and stiffer nails are in greater danger of having their value destroyed by the treatment accorded crates during handling and shipment.

Bolts and Bolting—Table 14 is from War Department Supply Circular No. 22, 1918, and it gives the relative size of bolts to be used in crate frames. There should be at least two bolts in each end of frame members. (See figures 1, 2, and 3, Plate XI.)

Carriage bolts are usually preferred for crating work because the heads are oval and do not catch on objects, as do

[1]See nail tables on pages 139, 140.
[2]See groups of woods, page 100.

the heads of machine bolts when not countersunk. Machine bolts usually require a washer under the head to give sufficient bearing surface. Carriage bolts also have the advantage of a square shoulder on the shank adjacent to the head, which prevents turning of the bolt when drawn into the wood. Washers, preferably standard cut, should be used under all nuts to prevent them from cutting into the wood. So far as

TABLE 14. SIZE OF BOLTS FOR CRATING

Thickness of frame lumber in inches	Diameter of bolt in inches
1 to 1 1/2	3/8
1 1/2 to 3	1/2
3 and over	5/8[1]

possible, all nuts should be put on the inner side of joints. Holes for bolts should be small enough for a drive fit, which will make a more rigid construction.

Lag Screws—Lag screws are not considered a good fastening in crating work. They may, however, be used should it be impossible to get a bolt into position and apply the nut, or where a bolt of excessive length would be required. In using lag screws, a hole should first be bored equal to the diameter and depth of the shank, and then the hole continued with a diameter equal to that at the root of the thread until the total depth of the hole is equal to the length from the head to the end of the untapered part of the thread

Straps—Straps may be used in strengthening crates. One method of strengthening a corner is shown in figure 7, Plate XI. Straps are also used in some instances to help keep the contents in position and support internal braces.

Binding Rods—Binding rods, or tie rods, may be run through crates in various directions to bind the parts more securely together. They are especially valuable in places where much tension is apt to be developed in the members and because, as has been mentioned, it is very difficult to join wooden members in such a way that their tensile strength can be utilized.

Internal Bracing—The object of internal bracing is primarily to support the contents in the crate in such a manner that the crate may ride on any face without injury to the

---

[1]In very heavy crates, bolts large than 5/8-inch will be desirable in some instances.

contents. If the contents are fairly rigid such internal bracing will then serve to strengthen the crate.

Internal bracing should, if possible, be so placed that the thrusts come against the end grain of the braces, then if moisture content of the braces is reduced they will not permit as much movement of the crate contents as would have occurred had the thrust been against the side grain of the braces, because shrinkage of wood parallel with the grain is negligible.

FACTORS DETERMINING AMOUNT OF STRENGTH REQUIRED IN CRATES

The weight of the contents and their resistance to external forces determine to a great extent the degree of strength that a crate must have. If the contents are heavy and rigid, possessing great strength, then a crate may be largely held in position and its shape maintained by internal braces[1] set at various places between the contents and the crate members. With contents which are more delicate and possess little strength, the crate must have much rigidity and strength so that it may be depended on to support the contents properly and prevent damage.

HAZARDS OF TRANSPORTATION

A crate must be strong enough to prevent damage to the contents from the hazards which may be encountered in its journey. Many American shippers at the present time have little conception of the severity of the hazards in export shipping.

In moving crates on rollers and handling with cranes severe bending stresses are apt to be produced.[2] Crates when being handled by cranes may also be dropped or collide with other objects in being lifted and swung, which will twist and strain them severely.

In moving crated material by wagon and trucks, the greatest hazards will ordinarily occur in the process of loading and unloading.

The hazards to be guarded against in shipment of crated materials over railroads on flat cars are movement of contents in the crate, movement of the crate on the car, theft of easily removed parts, and damage by the elements.

[1]See page 84
[2]See page 78.

In export shipment the hazards that are usually met with are very rough handling by cranes and derricks, the stresses that occur due to the method of stowage, cargo shifting, and the destructive action of the elements.

FACTORS INFLUENCING THE SIZE OF CRATES

The outside dimensions of a crate will depend on the dimensions of the contents, the amount of clearance necessary between the contents and the members, and the size of the members. As much disassembling of contents as is possible at a reasonable cost should be done, in order to reduce the space required.

The cost of material for making crates, storage space required, and charges for export shipment based on space occupied, are some of the factors which necessitate reducing the displacement of a crate to the minimum. With very large crates the ability of the transportation machinery to move them to their destination without difficulty is sometimes a very important consideration.

Before any large shipments are made, especially in foreign trade, complete information should be obtained. The National Association of Box Manufacturers is in touch with the various sources of information as to the many requirements for shipments to different foreign countries. These traffic requirements are so varied and voluminous that they cannot be included in this book.

CHAPTER IV

BOX AND CRATE TESTING

Methods of Testing and Their Significance

Data of value in the proper designing of boxes and crates cannot be easily obtained by observing boxes and crates as they proceed through the various stages of commercial service. Containers which have failed in service may be examined, but the causes which produced the failure can not be measured or readily observed, nor can the sequence of failures be told. Laboratory tests, however, which closely simulate the hazards encountered in commercial transportation service and which may be completely observed through all the stages of the life of a box or crate as it passes through the process of testing, give information as to the relative value of different details of construction. Laboratory tests can also be carried on more quickly and economically.

Some of the definite phases of box and crate construction work concerning which information may be obtained by testing are the following

1. Classification of woods as to nailing and strength properties for box construction.

2. Determination of balanced construction.[1]

3. Determination of the effect of different degrees of moisture content and changes in moisture content on the strength of boxes.

4. Comparison of various methods and amounts of reinforcing.

. 5. Comparison of different styles of construction.

6. Determination of the effect of various types of contents and methods of packing on the serviceability of a box.

7. Standardization of strapping of wooden boxes.

METHODS OF TESTING AND THEIR SIGNIFICANCE

There are now three types of box tests made at the Forest Products Laboratory: (1) drum, (2) drop, and (3) com

---

[1]See page 43.

FIG. 23—Testing boxes in small revolving drum developed at Forest Products Laboratory.

Fig. 24—Standard large drum testing machine developed at Forest Products Laboratory.

FIG. 25—Method of making drop-cornerwise test
A. Releasing device, or trip.   B. Cast iron plate.

pression. Drop tests may be made cornerwise or edgewise of the box; compression tests are made on the edges, corners, or faces. In general, all these tests lead to the same conclusions; the one selected in any particular case, however, will depend on the hazards to which the containers under consideration are subjected in transit.

<h2 align="center">DRUM TEST</h2>

The revolving drum type of box testing machine illustrated in figures 23 and 24 is an approved and very practical method of testing boxes and crates. A vertical section of the drum at right angles to the axis is hexagonal in shape, which gives it six faces. Upon these six faces, hazards and guides are arranged in such a manner that, as the drum revolves, the box or crate slides and falls striking on ends, sides, top, bottom, edges, and corners in such ways that the stresses, shocks, and rough handling of actual transportation conditions are simulated. For this method of testing the box or crate must be loaded with its contents or a substitute which produces the same effect.

The six faces, cleats, and edges of a box, numbered for testing in a revolving drum are shown in figures 3 and 4, Plate XIV; and the corners of a crate and other parts as numbered are shown by figures 1 and 2, Plate XIV. This numbering is necessary in order that a record of the locations and character of the failures may be made as they occur. As the box moves on from one drop to the next, the observer notes the beginning of any failure, and he follows the progress of that and any other failure until the box becomes unserviceable.

The weak feature of the box may be too few nails, nails of too short length, nails driven in a crack and thus having no great holding power, or some other form of nail weakness which the tests will clearly show. The material in the sides, top, or bottom may be too thin, so that the shocks of the falls pull the wood from the nails; the wood may split or break across the grain.

Any one of the numerous weaknesses of packing box construction may be demonstrated in this test, which enables the observer to design a box about equally serviceable in every feature, or balanced in construction. Such boxes will show failures equally likely to occur in nails pulling from the wood, wood pulling from the nails, splitting or breaking of ends, sides, tops or bottoms, and through the weaknesses of the wood species.

DROP TESTS

Drop-Cornerwise Test—In the drop-cornerwise test, a box or crate with its contents is suspended by each of its corners alternately and dropped from a definite height upon a cast iron plate or other solid surface, as illustrated in figure 25. This is a very good test for comparing the strength of various types as regards their ability to resist this particular hazard of sudden shock and distortive action. The test is very severe, however, and several failures are apt to occur simultaneously, so that the test is not as good as the drum test for drawing conclusions as to improvements of the design.

In the drop test the corners of the box or crate where the various faces meet should be numbered as follows:

Faces meeting	Corner No.
5-1-2	1
6-3-4	2
5-2-3	3
6-1-4	4
5-3-4	5
6-1-2	6
5-1-4	7
6-2-3	8

The box or crate should then be dropped on the corners in numerical rotation and then the cycle repeated until failure occurs. The height of the drop is usually increased for each repetition of the cycle.

Drop-Edgewise Test—The drop-edgewise test is similar to the drop-cornerwise test except that the container is dropped on an edge instead of a corner, being suspended from the diagonally opposite edge.

COMPRESSION TESTS

Compression-On-An-Edge Test—In the test illustrated by figure 26, a compressive force is exerted over the whole length of one edge of a box and at right angles to it, the direction of the force also passing through the diagonally opposite edge. This test measures the ability of a box or crate to resist being collapsed in this particular manner by external forces, enables comparisons to be made of the relative resistance in containers of the same design, and provides an additional method of comparing the strength of containers of different

Fig. 26—Method of making compression-on-an-edge test.

Fig. 27—Method of making compression-cornerwise test.

FIG. 28—Method of making compression-on-faces test.

designs. This test is usually applied to the empty box or crate, although it may be applied to the container as packed for shipment.

Compression-Cornerwise Test—Another compression test, figure 27, is conducted by applying compressive forces to two corners of a box or crate and directly along the line passing diagonally through those corners and the center of the box or crate. The general purposes of such a test are similar to those given for the preceding test, although it has the advantage of more readily determining the weakest elements of the construction. Each of these tests, however, brings out certain strength factors which the other does not, so that one can not be substituted for the other.

Compression-On-Faces Test—A test which subjects a box or crate to the stresses that it encounters when supporting heavy static loads in storage warehouses is obtained by direct compression at right angles to any two parallel faces, as illustrated in figure 28.

Supplementary Tests

In addition to tests on completed boxes and crates there are many tests that will give much information of value to the designer, manufacturer, and user of boxes. Among these tests are the following:

1. Mechanical-properties tests on sawed lumber, rotary-cut lumber, and plywood. (See page 26.)

2. Holding power of nails, screws, bolts, and other fastenings.

3. Tensile strength tests on metal strapping and wire ties and various methods of fastening them.

4. Density determinations for woods to identify weak and brash stock.

5. Determinations of the percentage of moisture in wood.

6. Strength tests on glued joints

7. Tests on the strengthening effects of corrugated fasteners.

8. Tests on special materials and details, such as rope, webbing and metal handles, hinges, hoops, locks, etc.

As new designs of containers and their various accessories are developed, corresponding tests will be necessary to determine their strength and advantages as compared to other containers and devices of a similar character.

CHAPTER V

BOX AND CRATE SPECIFICATIONS

STANDARDIZATION OF PACKING BOXES—NATIONAL ASSOCIATION OF BOX MANUFACTURERS' TENTATIVE GENERAL SPECIFICATIONS FOR NAILED AND LOCK CORNER BOXES—SPECIFIC SPECIFICATIONS FOR NAILED AND LOCK CORNER BOXES—GENERAL SPECIFICATIONS FOR 4-ONE AND SIMILAR BOXES.

Purpose—The purpose of box or crate specifications is to provide that part of a contract or agreement usually made between a box manufacturer and his customer or other contracting parties which sets forth all of the details pertaining to the materials used and the construction of the container as they are to be furnished or executed by the manufacturer. The statements in a specification should be definite, concise, and clear.

The specifications for a box or crate, so far as they affect its actual strength, should only be such as will enable the container to perform satisfactory service. To require more than this will usually make the containers less economical.

General specifications for wooden boxes nailed and lock corner are given below. They are divided into four parts, viz., material, grouping of woods, dimensions of parts, and manufacture.

These general specifications are intended to serve as master specifications or standards of construction for nailed and lock-corner boxes. It is expected that specifications for boxes for a specific commodity or group of commodities will, as they are worked out through careful research, be made to conform closely to these general specifications, and that many of their requirements will be specified by reference to the general specifications. •It will be necessary to add only such items as style and dimensions of box and minimum thicknesses of parts, and to enumerate such exceptions to the general specifications as may be found necessary. Many points must of necessity be common to all boxes of the classes under consideration, and brevity is gained by the scheme of specifying

97

these in standards for boxes for a specific commodity by reference.

The general specifications have the further value of providing a much needed standard as a guide to the construction of wooden boxes. The principal features of these specifications have been worked out from a large number of careful tests of boxes as such, and from very extensive data on the mechanical properties of woods as determined by the Forest Products Laboratory

STANDARDIZATION OF PACKING BOXES

Paper Presented to National Association of Box Manufac turers in Convention, April, 1920

"Standardization of packing boxes for any commodity, like standardization of other products, tends to increase production, to insure uniform quality and to lower costs. This applies not only to uniformity of dimensions of the box and its parts, but more especially to all those specifications of quality which directly affect the strength properties of the package.

"Processes of standardization require system and order first of all. Haphazard assembling and grouping of essential specifications lead to duplications, errors, and omissions and tend to destroy the effectiveness of the work. No attempt will be made to complete this project before distributing the work. Instead, the general specifications and probably several chapters of detailed specifications will be distributed as compiled; succeeding chapters to follow as they are completed. It is advisable that those who receive these various installments should maintain a file for them.

"This project of standardization for nailed and lock corner wooden boxes is based on a definite, systematic program, which will set forth, logically, concisely and completely, the best that can be compiled by this association, from its practical experience, combining with this all that has been developed in the scientific studies by trained engineers in the laboratory.   •

"As in all other types or kinds of packing boxes, certain fundamental specifications are common to all nailed and lock corner wooden boxes. It is right and proper that these fundamentals be stated as the first chapter in this work of standardization. In each succeeding chapter, which deals with the detailed specifications that apply to its particular group of boxes, reference will be made to the general specifications only when an exception is to be made in the interest of greater efficiency or lower costs of production.

"Each chapter or bulletin will be devoted to the boxes in common use for one general class of commodities, giving the specific thicknesses of material to be used for those particular boxes, and where practicable preferential sizes, together with such comments and notations as may be desirable in the interest of proper packing.

"In compiling these specifications, acknowledgment is made for the careful and comprehensive work of the U. S. Forest Products Laboratory, Madison, Wis., and for the co-operation of those leaders in the box making industry whose experience has been drawn upon to make these specifications thoroughly practicable. The continued work of research will produce new and greater economies, this, with changing manufacturing and commercial conditions, will necessitate frequent changes in standardized box construction and in specifications.

"To those associations of shippers and to those in transportation lines, whose activities require them to compile standard of box construction, whether for wooden boxes or for any of the other types or kinds of containers, this method of compiling, (1) the general specifications and (2) the detailed specifications, is recommended as absolutely necessary if clear, concise, comprehensive, efficient and economical results are to be obtained.

NATIONAL ASSOCIATION OF BOX MANUFACTURERS' TENTATIVE GENERAL SPECIFICATIONS FOR NAILED AND LOCK CORNER BOXES[1]

A. MATERIAL

Material—The ends, sides, tops, bottoms and other parts of wooden boxes must be well manufactured and be cut true to size. All defects in the lumber that materially lessen the strength of the part, expose contents to damage, or interfere with proper nailing, must be eliminated. The lumber must be thoroughly seasoned, viz., have an average moisture content of 12 to 18 per cent, based on the weight of the wood after oven drying to a constant weight.

---

[1]Adopted as tentative general specifications for nailed and lock corner boxes by the American Society of Testing Materials.

B. Grouping of Woods

Grouping of Woods—The principal woods used for boxes are classed for the purpose of specifications in four groups:

Group I

White pine	Willow
Norway pine	Noble fir
Aspen (popple)	Magnolia
Spruce	Buckeye
Western (yellow) pine	White fir
Cottonwood	Cedar
Yellow poplar	Redwood
Balsam fir	Butternut
Chestnut	Cucumber
Sugar pine	Alpine fir
Cypress	Lodgepole pine
Basswood	Jack pine

Group II

Southern yellow pine	Douglas fir
Hemlock .	Larch (tamarack)
North Carolina pine	

Group III

White elm	Black ash
Red gum	Black gum
Sycamore	Tupelo
Pumpkin ash	Maple, soft or silver

Group IV

Hard maple	Birch
Beech	Rock elm
Oak	White ash
Hackberry	Hickory

C. Thickness of Lumber

Thickness of Parts—The thicknesses called for in specifications for boxes of any given commodity will, unless otherwise stated, be understood as applying to Groups I and II woods. Where the material is specified (for Groups I and II woods) as not more than $\frac{1}{2}$ inch thick and not less than $\frac{3}{8}$ inch, Groups III and IV woods may be used $\frac{1}{16}$ inch less in thickness; where the material is specified (for Groups I and II woods) as more than ½ inch thick and not more than 1 inch, Groups III and IV woods may be used ⅛ inch less in thickness; where the material is specified (for Groups I and II woods) as more than 1 inch thick and not more than 2 inches, Groups III and IV woods may be used ¼ inch less in thickness.

The thickness of lumber specified allows for an occasional unavoidable variation, but that variation shall not exceed one-eighth of the thickness of the part below the thickness specified.

D. Widths of Material

Widths of Parts—The maximum number of pieces allowed in any side, top, bottom, or end of a box shall be as follows:

Width of face	Maximum number of pieces
5 inches or under	1
Over 5 or under 8 inches, inclusive	2
" 8 " " 12 " "	3
" 12 " " 20 " "	4

For each additional 5 inches in width of face, one additional piece may be used. No piece less than 2½ inches face width at either end may be used in any part, except for cleats or battens.

E. Surfacing

Surfacing—The outside surfaces of boxes must be sufficiently smooth to permit of legible marking.

F. Joining

Joining—Ends 1 inch or less in thickness, if made of two or more pieces, must be either butt-jointed or matched, then fastened with two or more corrugated fasteners or must be cleated. For ends ⅞, 1¾₆, and ¾ inch thick, use fasteners 1 by ⅝ inch; for ends ⅝ inch thick, use fasteners 1 by ½ inch; for ends ½ and ₁₇₆ inch thick, use fasteners 1 by ⅜ inch. Two or more pieces Linderman jointed shall be considered as one piece.

G. Schedule of Nailing

Size of Nails—All nails specified are standard cement coated box nails. (If other than cement coated nails are used 25 per cent more nails must be driven than specified). Plain nails driven through and clinched may be used for cleating.

The size of the nail to be used shall be governed by the species and thickness of the material in which the points of the nails are held. If the designated penny of nail is not available, use the next lower penny and space nails proportionately closer. Nails should be driven flush—overdriving materially weakens the container.

Nailing Schedule

Use cement coated nails of size indicated when species of wood holding nails is	Thickness of ends or cleats to which sides, tops and bottoms are nailed								Thickness of sides to which top and bottom are nailed		
	3/8 or less	7/16	1/2	9/16	5/8	11/16 or 3/4	13/16	7/8	Less Than 1/2	1/2 to 9/16	5/8 to 7/8
Group I woods	4d	5d	5d	6d	7d	8d	8d	9d	4d	6d	7d
Group II woods	4d	4d	5d	5d	6d	7d	7d	8d	4d	5d	6d
Group III woods	3d	4d	4d	5d	5d	6d	7d	7d	3d	4d	5d
Group IV woods	3d	3d	4d	4d	4d	5d	6d	7d	3d	4d	5d

H. Spacing of Nails

Spacing of Nails—In order to ascertain the number of nails to be used, divide the width of the side, top and bottom (or length of cleat) by the spacing specified for the gauge of nails to be used. Fractions in the result greater than $\frac{1}{4}$ (if the points of nails are to be held in the end grain) and greater than $\frac{1}{2}$ (if the points of nails are to be held in the side grain) will be considered as a whole number.

	Space when driven into	
When nails are	Side grain of end	End grain of end
6d or less	2 inches	1¾ inches
7d	2¼ "	2 "
8d	2½ "	2¼
9d	2¾	2½
10d	3	2¾

No board shall have less than two nails at each nailing end. Where cleats of thickness not less than the thickness of the ends are used, approximately 50 per cent of the nails will be driven into the cleats.

Space nails holding top and bottom to sides six to eight inches apart. (When material in sides is less than $\frac{1}{2}$ inch thick, do not side nail unless otherwise specified.)

SPECIFIC SPECIFICATIONS FOR NAILED AND LOCK CORNER BOXES[1]

The following are specifications for Nailed and Lock Corner boxes for carrying specific commodities, minimum thickness of lumber and maximum gross weights

CANNED FOOD CASES

Commodity	Type of box construction	Minimum thickness of the material expressed in fractions of an inch, woods of Groups I and II. (See General Specifications for comparative thicknesses of woods in Groups III and IV)			Maximum gross weight of box and contents	For exceptions to general specifications see reference to notes as numbered
		Ends	Sides	Top and bottom		
Canned f o o d s in metal cans, viz.:	Nailed..	5/8	5/16	5/16	90	Notes 1, 2
		5/8	3/8	3/8	125	Notes 1, 2, 3
fish, fruits, meats, milk, vegetables, other foods.......	Lock-corner.	7/16	7/16	5/16	50	Notes 1, 2
		1/2	5/16	5/16	50	Notes 1, 2
		5/8	5/16	5/16	90	Notes 1, 2

NOTES AND EXCEPTIONS

Note 1—When one-piece sides and two-piece tops and bottoms of Groups I and II woods are used, material may be $\frac{1}{32}$ inch thinner than specified. When rotary cut gum lumber is used, with one-piece sides and tops and not more than two piece bottom, the thickness may be $\frac{1}{16}$ inch less than specified for Group III woods, minimum thickness $\frac{1}{4}$ inch.

Note 2—Inside dimensions of boxes shall not exceed $\frac{1}{4}$ inch over exact length and width and $\frac{1}{8}$ inch over exact depth of contents.

Note 3—Ends *must* be cleated.

GENERAL SPECIFICATIONS FOR 4-ONE AND SIMILAR BOXES

The following specifications are substantially those adopted as general specifications by the 4-One Association of Box Manufacturers, and as tentative general specifications by the American Society for Testing Materials

---

[1]The foregoing general specifications would be incomplete without an illustration of their application to a specific commodity. Therefore, specific specifications for canned food containers, as adopted by the National Association of Box Manufacturers, are set forth. Specifications for boxes for other commodities have been adopted, but will not be incorporated in this book.

1. These specifications cover three styles of 4-One and similar type boxes as follows:

(a) General form..
(b) Boxes with wedgelock ends.
(c) Boxes with detached tops.

2. **General Form**—(a) The boxes knocked down shall consist of four separate sections forming top, side, bottom, side, connected only by continuous steel binding wires; and of separate ends.

(b) Each of the separate sections forming the sides, top, and bottom shall consist of cleats, thin boards, wires, and staples.

(c) The four sections shall be separated such a distance from each other that the wires shall be in tension at the corners when the sections are folded.

3. **Grouping of Woods**—For the purposes of these specifications, box lumber shall be classed into four groups as follows:

Group I.

Alpin fir	Cottonwood	Redwood
Aspen (popple)	Cucumber	Spruce
Balsam fir	Cypress	Sugar pine
Basswood	Jack pine	Western yellow pine
Buckeye	Lodgepole pine	White fir
Butternut	Magnolia	White pine
Cedar	Noble fir ·	Willow
Chestnut	Norway pine	Yellow poplar

Group II.

Douglas fir	Southern yellow	Virginia and
Hemlock	pine	Carolina pine
Larch		
(tamarack)		

Group III.

Black ash	Red gum	Sycamore
Black gum	Red gum sapwood	Tupelo
Maple, soft or silver	(commonly	White elm
Pumpkin ash	called sap gum)	

Group IV.

Beech	Hickory	Rock elm
Birch	Maple, hard	White ash
Hackberry	Oak	

MATERIALS

4. **Cleats**—(a) Each cleat shall be sound, free from knots and from cross grain which runs across it within a distance equal to one-half its length

(b) Cleats shall be not less than $\frac{3}{4}$ inch thick (parallel to the length of the box) and not less than $\frac{7}{8}$ inch in width.

5. **Thin Boards**—(a) The thin boards shall be sound (free from decay and dote), well seasoned and cut so that adjacent faces of boxes will be at right angles to each other. All defects that would materially lessen the strength, expose the contents of the boxes to damage, or interfere with the proper assembly of the boxes shall be eliminated.

(b) When the thickness of thin boards as specified is less than $\frac{3}{16}$ inch, thin boards made of woods of Groups III and IV may be $\frac{1}{32}$ inch less than the specified thickness except that the minimum thickness of thin boards of any kind of wood shall be $\frac{1}{8}$ inch.

(c) The variation in thickness of thin boards below the thickness specified shall be not more than $\frac{1}{8}$ of the thickness of the thin board, and this variation below the specified thickness shall not extend to more than 10 per cent of the face of that particular board.

(d) Thin boards less than $2\frac{1}{2}$ inches in width at either end shall not be used.

6. **Staples**—The binding wires shall be annealed steel wire of not less than No. 16 gauge.

7. (a) The staples on end wires shall be not less than No. 16 gauge by $1\frac{1}{8}$ inches long.

(b) Staples on intermediate wires shall be not less than No. 18 gauge by $\frac{7}{16}$ inch long.

ASSEMBLING

8. (a) The staples on end wires shall be driven home astride the binding wires, through the thin boards into the cleats, and anchored in the cleats.

(b) The staples on the intermediate wires shall be

driven astride the binding wires, through the thin boards and firmly clinched.

(c) The space between staples shall be the average distance between centers of staples astride each binding wire in each section and this space shall be not more than 2½ inches except as specified in paragraph (d).

(d) When cleats are made of woods of Groups III and IV the space between the staples may be ¼ inch greater than that specified.

(e) There shall be not less than two staples driven astride each wire and into each thin board.

(f) The staples nearest the corners shall be not more than 1¾ inches from the corner to which it is adjacent.

9. Each end of the box shall be securely fastened on the inside of the side cleats with staples not less than No. 16 gauge by $\frac{13}{16}$ inch long, or with cement-coated nails of not less than two-penny size. There shall be no space exceeding 2½ inches on any side cleat into which no staple or nail holding the end in place has been driven and there shall be a staple or nail within 1½ inches of each end of each side cleat. Staples or nails shall be driven home.

10. At each corner one section shall overlap its adjacent section at right angles and the wire shall be in tension, giving a square, tight corner

11. The cover shall be closed tightly and the ends of each binding wire twisted tightly together. The twisted portion of each wire shall be not less than ½ inch long. The rough ends of the wires shall be removed and the twisted portion driven flat against the side parallel with the binding wire.

12. Nothing herein contained shall be construed as prohibiting the use of boxes constructed of thicker thin boards, additional or heavier wires, heavier cleats, longer staples, or with closer spacing of staples.

13. **Boxes with Wedgelock Ends**—These boxes shall consist of sides, top, and bottom and one end made in accordance with sections 2-12 inclusive, pages 104 to 106, inclusive, and of one wedgelock end.

14. Wedgelock ends shall consist of the following:

(a) One or more thin boards whose thickness is not less than that of the thin boards in the other portions of the box and whose combined width is ⅛ inch less than the shortest distance between the top and bottom cleats of the box and whose length is the same as the inside width of the box less the width of one of the side cleats.

(b) Two battens of the same thickness and width as the cleats in the box and whose length is ⅛ inch less than the shortest distance between the top and bottom cleats.

(c) One wedge of the same thickness and length as the battens and whose width is one-half that of the battens.

15. The battens shall be attached across the grain of the thin boards, one batten its own width from one end and the other batten half its width from the other end of the thin boards, with staples not less than No. 16 gauge by 1¾ inch long or with nails not less than two-penny size. There shall be no space exceeding 2 inches on any batten into which no staple or nail holding the thin boards to the batten has been driven and there shall be a staple or nail within 1½ inches of each end of each. batten. Staples or nails shall be driven home.

In making up the box, the wedgelock end is left out so that the box may be filled from the end. The other end of the box is fastened in place and the box made up as specified in sections 9-11, page 106. The wedgelock end is not fastened in place until the box is closed.

Boxes with wedgelock ends are closed as follows·

The wedgelock end is inserted. The wedge is inserted, the batten that rests against the wedge is fastened to the wedge with one four-penny nail driven through the middle of the batten into the wedge. The other batten that rests against the cleat is fastened to that cleat with four-penny nails driven through the batten into the cleat. Nail centers shall be not more than 4 inches apart, and there shall be a nail within at least 2 inches of each end of this batten.

16. **Boxes** with Detached **Tops**—These boxes shall consist of sides, bottom and ends made in accordance with sections 2-12 inclusive, pages 104 to 106, and a detached top made of thin boards.

17. In assembling the box, the top cleats to which the binding wires have been stapled shall be put in position on the side cleats and the ends of each wire stapled to the cleats twisted tightly together.

The detached top shall be nailed to the end cleats with cement-coated box nails spaced not more than 2½ inches apart. The wires not stapled to the cleats shall be brought over the detached top and the ends of each wire twisted tightly together.

CHAPTER VI

STRUCTURE AND IDENTIFICATION OF WOODS

STRUCTURE OF WOOD—PROCEDURE IN IDENTIFYING WOOD—KEY FOR THE IDENTIFICATION OF WOODS USED FOR BOX AND CRATE CONSTRUCTION—DESCRIPTION OF BOX WOOD—GRADING RULES FOR ROTARY-CUT LUMBER.

More than forty different species of wood are used in box construction. It is essential to be able to distinguish these woods in order that they may be used intelligently. For instance, it is necessary to be able to classify the commercial woods used for boxes into the four groups outlined in the specifications for boxes and crates. (See grouped list, Part V, page 100.) Such properties as color, odor, taste and weight are very helpful in placing any given wood in the group where it belongs; information as to the section of the country from which the wood was obtained is also of assistance; but some knowledge of structure is indispensable for accurately making the required distinctions. When a wood is dry it may lose much of the odor that distinguished it when green; if it is stained, weathered, or artificially treated, any characteristic color may not be apparent; its weight, too, when it is green (saturated with moisture), when partly seasoned, and when kiln-dried is very different; under all these conditions, however, the structure is practically unchanged and, therefore, serves as an unfailing guide in identifying the wood. Moreover, accurate descriptions of characteristic structures are possible, whereas descriptions of color and so-called "grain" are difficult to put into words and are open to wrong interpretation because of the variation of individual opinions and observations on "grain" and color.

The identification of woods as described in the following pages, therefore, is based primarily on the structure of the wood, but is supplemented by such other physical properties as are helpful in distinguishing the different species or groups of woods

STRUCTURE OF WOOD

HEARTWOOD AND SAPWOOD

In mature trees two portions of the wood of the trunk are generally to be distinguished. These are the sapwood and the heartwood. The sapwood is found next to the bark; it is generally light colored and varies only slightly in shade. It varies considerably, however, in width. In some species it is less than an inch wide, as, for example, in arborvitæ, western red cedar, black ash, and slippery elm. On the other hand, in other species, such as maple, birch, hickory, white ash, green ash, hackberry, and some hard pines, it is several inches in width. Besides the variations in sapwood in species, the width of sapwood may also vary within the same tree or species, depending upon the age, vigor of growth, and height above the ground of the individual specimen. The sapwood contains living cells and it is through this portion of the woody cylinder that the sap circulates in the tree.

The heartwood is dead so far as the life processes of the tree are concerned. The heartwood was once sapwood. After serving for sap conduction and other growth activities for a number of years the sapwood gradually, often without any sharp line of demarkation, changes into heartwood and ceases to function in the life of the tree except for the fact that it gives mechanical support to the crown, thus helping hold the leaves up in the sunlight. The change from sapwood to heartwood is very often, but not always, accompanied by a change in color. Some woods in which the change in color is lacking or very slight are white and red spruce, hemlock, Port Orford cedar, basswood, white cottonwood, aspen or "popple," buckeye, "whiteheart" beech[1], and hackberry. The color of the heartwood, when markedly different from that of the sap wood, is of great assistance in identifying different species of wood, such as red gum, yellow poplar, and black ash.

The structure of the sapwood and the heartwood is the same except for the fact that the pores or large tube-like cells of the heartwood of some hardwoods are frequently more or less closed with cell-like growths called tyloses or with gummy substances. In the sapwood, especially in the outer sapwood next the bark, the pores are open and serve to conduct sap. Sometimes the sapwood of lumber becomes much discolored, blued, or darkened, through the presence of sap-

[1] Some beech has a very reddish heartwood frequently with different properties from the "whiteheart" beech—this is often called "redheart" beech.

stain fungi. Some woods in which the sapwood is often blued or otherwise stained are pines, spruces, red gum, and hackberry. In hackberry the stained sapwood often appears darker in color than the heartwood. Discoloration may also be produced by chemical changes in the wood without the action of fungi; for example, brown stain in sugar pine or the colors produced by the contact of the saw with the substances in oak.

ANNUAL RINGS

Annual rings are the more or less well defined concentric layers of wood laid down each year by the growing tree. They are particularly noticeable on the stump of a tree or on any cross section of the wood. Annual rings are more conspicuous in oak, elm, and ash, or pine and fir than they are in woods like "popple" or aspen, buckeye, cottonwood, tupelo, and willow. Many woods grown in the tropics do not show well defined annual rings, although they may show zones of growth due to changes, often not annual but produced by climatic conditions other than the summer and winter changes of tem perate climates.

The appearance of the annual rings on a smoothly-cut cross section is of great assistance in identifying woods.

SPRINGWOOD AND SUMMERWOOD

The springwood is that part of the annual ring which is first formed each year. The wood is usually lighter in weight and softer because it contains more air spaces, that is, larger cell cavities and less wood substance than that formed later on in the year. As the season advances the cell walls formed are usually thicker and the cavities smaller so that the growth which is called summerwood is denser, harder, and often darker in color than the springwood. In some woods, such as maple, birch, and basswood, there is a little difference between the springwood and summerwood. In the case of spruce or hemlock the change from springwood to summerwood is gradual, but in oak and longleaf pine, for instance, the difference between springwood and summerwood is not only marked but the change is very abrupt. It is often possible to estimate the strength of wood by noting the percentage and density of the summerwood.

FIG. 29—Section of western yellow pine log showing: radial surface, R; tangential surface, T; heartwood, H; sapwood, S; pith, P. The annual rings are the concentric layers widest near pith and usually becoming narrower toward the bark. The summerwood produces the dark lines that stand out conspicuously on both the radial and tangential surfaces.

The Structure of Hardwoods

The name "hardwood" applies chiefly to woods which are characteristically hard and strong, such as oak, hickory and ash. The hardwood group, however, includes some woods which are not relatively very hard, as, for instance, basswood and "popple." The real distinction on which the grouping into hardwoods and softwoods is based is not the hardness but the structure of the wood.

All the commercial hardwoods of the United States contain pores or vessels, cells which are strikingly larger than the other cells with which they are associated. The conifers or softwoods do not have these pores or vessels. They are often called non-porous woods. Some of them, for example southern yellow pine or Douglas fir, may be actually harder than such hardwoods as basswood. The hardwoods, for the most part, come from broad-leaved trees, such as maple, elm, and poplar, and the softwoods from needle or scale-leaved trees like pines or cedars.

All hardwoods have pores. In many cases pores are visible to the naked eye. They appear on a smoothly-cut cross section as more or less circular openings. On a longitudinal section they appear like fine, more or less interrupted grooves and may be used to determine the direction of the grain, as, for example, cross or spiral grain. Where they are not visible to the naked eye they may be seen with the aid of a magnifying glass with an enlarging power of about 12 to 18 diameters. The group, however, may be divided into two classes according to the arrangement of these pores. When the large pores are grouped conspicuously at the beginning of each annual ring and there is an abrupt change in size from the springwood to the summerwood pores the woods are called ring-porous. The springwood pores are usually visible to the naked eye in this type of wood. Examples of ring-porous woods are oak, ash, elm, hickory, and locust. In this type of wood the annual rings are distinctly defined.

If, on the other hand, the pores are scattered with considerable uniformity throughout the ring and the change in the size of the pores from the inner to the outer portion (spring to the summerwood) is slight or gradual, the woods are called diffuse-porous. Examples of woods of this sort are maple, birch, tulip, gum, sycamore, and willow.

Tyloses are cell-like growths which often appear like a

froth or a number of glistening particles in the pores of the hardwoods. They are especially conspicuous in such woods as hickory and most of the white oaks. Tyloses, when present, are found in the inner sapwood and the heartwood. The outer sapwood pores are normally open. The presence of tyloses is often of assistance in identifying woods. White oak with an abundance of tyloses, for instance, is used for tight cooperage (barrels to contain liquids), while red oak, which generally (not always) lacks tyloses, is not as suitable for liquid containers and is used for slack cooperage (barrels for dry materials such as cement or flour).

Lines of light colored tissue extending out from the pores may be seen on the smoothly-cut cross section of such woods as white, green, or pumpkin ash. These lines are of considerable assistance in identifying these woods. They are composed of small rather thin-walled cells which may easily be crushed when a section is cut. These cells are known as parenchyma tissue.

Rays (often called medullary rays) are more or less narrow strips of cells which extend from the bark towards the center of the tree. They run horizontally at right angles to the vertical grain of the wood. They may be compared to minute two-edged swords thrust from the inner bark toward the heart of the tree. On the end surface they appear as lines, crossing the annual rings, like the radii of a circle or the spokes of a wheel, although all the rays do not originate at the center of the tree. The rays are very large and conspicuous in oaks, in which they are sometimes said to produce "silver grain" or "fleck." The rays are also easily seen with the naked eye in sycamore and beech. Although they are visible in all woods on truly radial surfaces, especially on split surfaces, and in many woods on the end surface there are also a considerable number of species in which they cannot be seen on the end surface with the naked eye. Quarter-sawed or edge-grain material is produced by cutting through the center of the tree approximately parallel to these rays (radial cut). Flat-grained, slab, or plain-sawed material is cut at right angles to the rays (tangential cut)

Wood fibers are the small cells which make up the greater part of the dense wood substance between the pores and the rays of hardwoods. They are thick-walled and too small to be seen individually without considerable magnification. Collectively, they are seen to form the darker, denser portions which give most of the weight and strength to the wood.

Pith flecks are small dark spots or streaks which occur characteristically in certain woods, as, for example, soft maple and river birch.

The Structure of Conifers

In conifers or softwoods the rays and the fiber-like cells (tracheids) make up the wood. The tracheids serve the combined purpose of the pores and the wood fibers of the hardwoods, that is, they support the tree and assist in the conduction of the sap. The truth of the statement that wood resembles a honeycomb is strikingly evident when the structure of a softwood is examined under a lens. In the wood the cells are, in proportion to their width, longer than they are in the honeycomb, although they are rarely over ¼ inch in length. The resemblance is due to the regularity of arrangement of the cells, which are of approximately uniform width tangentially and are arranged in very regular radial rows, as is apparent in the pictures of conifer cross sections. (See figure 2, Plate II.) Because the conifers do not have cells which are strikingly larger than the other cells (pores) they are called non-porous woods or woods without pores. (It should be noted that the word "porous" as applied to substances like a sponge, which contain empty spaces and may absorb liquids, may be applied also to both softwoods and hardwoods which, of course, contain air spaces; but the word "pore" or "vessel" is used in the classification of hardwoods and conifers in a different sense when reference is made to the presence of a definite type of cell in the hardwoods or "porous" woods.)

In the softwoods the annual rings are usually very clearly defined (more clearly than in some diffuse-porous woods) because toward the close of the growing season the cells become thicker walled and are flattened somewhat radially, thus producing the distinctive summerwood of the annual rings.

The rays in the conifers, although present, are so small as to be invisible on the end surface without a lens. They are very numerous, however, as many as fifteen thousand to the square inch (21 to 27 per square mm.) of tangential surface have been noted in such a wood as pine.[1]

---

[1] U. S. Department of Agriculture, Division of Forestry, Bulletin 13. The Timber Pines of the Southern U. S., page 152.

Resin passages or ducts are found in four genera of the conifers, namely, in pines, spruces, Douglas fir, and larch or tamarack. Resin ducts are openings which have been produced when the cells of the wood have split apart, that is, they are intercellular spaces. These passages or ducts extend both vertically and horizontally in the tree. The vertical ducts in the woods mentioned usually occur in or near the summerwood and are more visible than the horizontal ducts which are found in some of the rays (fusiform rays). These latter may be seen as very minute dark specks on the tangential surface of the woods of the species just mentioned. Resin is stored in these ducts or intercellular spaces and often gives them a brownish or amber color which assists in their detection, especially on longitudinal surfaces. The ducts, especially the vertical ones, are most conspicuous in pines. When a very smoothly-cut end-grain surface is examined the vertical resin ducts are barely visible to the naked eye as minute dots in or near the summerwood. In spruces, larches, and Douglas fir the resin ducts are usually smaller and less numerous than in pines, and they are sometimes found in short tangential rows. The direction of the grain, especially spiral grain, may be determined by observing the vertical resin ducts which run parallel with the fibers. Exudation of resin sometimes occurs from the ducts on the ends of pieces of these four kinds of woods. The absence of such exudations of resin does not necessarily mean that resin ducts are not present, for, as a rule, the resin does not exude from the ducts in seasoned wood unless the wood is heated. In woods in which resin ducts are present pitch pockets or pitch streaks may also be found. Such coniferous woods as cedars, cypress, redwood, and balsam firs do not normally have resin ducts. They may, however, contain some resinous material in their rays or in certain scattered cells.

PROCEDURE IN IDENTIFYING WOOD

If color, odor, weight, or general appearance is not sufficiently distinctive to identify a specimen of wood, an examination of the more detailed structure should be made. In the key which follows, the somewhat similar woods are systematically grouped together and the characteristic distinctions by which they can be separated are given to assist in rapid and accurate identification. Photographs showing a slightly magnified cross section of the different species are also given. These show very clearly the distinctions which may usually also be seen with the naked eye on a smoothly-cut surface of the end grain of the wood. Contrary to common practice, it is the cross section or end grain, which, when smoothed off, with a very sharp knife, usually presents the best surface from which to make an identification; for, in general, it is here rather than on the longitudinal surface that the principal distinctive characteristics in a difficult identification are to be found. The need of a very sharp knife and smoothly-cut surface cannot be too strongly emphasized. With a dull knife scratches and other irregularities may be produced which are sometimes mistaken for structures.

To determine the color of a wood a freshly-cut longitudinal surface of the heartwood should be used, since exposed surfaces may become weathered or soiled so that the characteristic color is changed. Odor and taste should also be determined from freshly-cut surfaces, shavings, or sawdust, since they are often lost if the material is exposed to the air for any length of time.

The first step in identifying an unknown wood is to determine whether or not pores are present. In many woods the pores are readily visible to the naked eye, but in some they are difficult to see or invisible without magnification. This is particularly true in certain diffuse-porous woods, such as sycamore, beech, red gum, maple, yellow poplar, basswood, tupelo, buckeye and aspen. In the case of practically all of these woods, however, there are characteristics which readily distinguish them from the somewhat similar appearing softwoods where pores are lacking. These are, for instance, the relatively large rays found in beech, sycamore, maple, and basswood, the color of the heartwood of red gum and yellow poplar, and the lack of sharply-defined annual rings in tupelo, aspen, and buckeye, together with other details which are given in the key, thus making it possible readily to separate these hardwoods from certain of the softwoods for which they might be mistaken.

After determining whether a wood is a hardwood (with pores) or a softwood (without pores), the next step is to place the wood under one of the sub-divisions under the group to which it belongs; that is, if it is a hardwood, note whether it is ring-porous or diffuse-porous. Then if it is a ring-porous wood, note whether or not the summerwood is figured and, if so, whether the figure of the summerwood runs with the rays across the rings (radial) figures 1, 2, and 3, Plate XVI, or with the rings (tangential), figures 4, 5, 6, 7, and 8, Plate XVI.

The most distinctive features within the diffuse-porous group are the size of the pores and the size of the rays, the color and the weight of the wood

If, on the other hand, the wood is a conifer (without pores), the annual rings are usually well defined by the contrast between springwood and summerwood. The principal characteristics in this group are odor, presence of resin ducts in certain woods, color, and weight.

It is not to be expected that the key can be used successfully without some practice. It is also very desirable that the person who is to make a specialty of wood identification should have a collection of known samples of wood which show characteristic color and structure (that is, not extremely fast or extremely slow growth) for comparison. It should be noted that in the key the name of the wood *follows* its description.

KEY FOR THE IDENTIFICATION OF WOODS USED FOR BOX AND CRATE CONSTRUCTION[1]
(without the aid of hand lens)

HARDWOODS

Woods from broad-leaved trees.

Woods with **pores** or **vessels,** that is, cells larger than those surrounding them.

I. **Pores** present—(Sometimes not visible to the naked eye in certain diffuse-porous woods, in which, however, the distinct rays or lack of well-defined summerwood distinguish them from conifers.)

    A. Ring-porous woods—The comparatively large springwood pores are clearly visible, especially in the sapwood at the beginning of each annual ring. On the end grain of a log these pores form distinct rings. The marked difference between springwood

---

[1]Unless otherwise stated all observations of structure are made on a smoothly cut cross section or **end grain** showing growth rings of average width. A sharp knife is indispensable. All **color** determination should be made on a freshly-cut longitudinal surface of the heartwood. See pages 126 to 136 for a more detailed discussion of each wood.

and summerwood is characteristic. Longitudinal surfaces appear **coarse textured** because of the large springwood pores which show as fine grooves or furrows, often producing a characteristic figure. These woods are mostly heavy and are found in box wood groups three and four. Chestnut, which is a ring-porous wood, is an exception; it is fairly light when seasoned and is classified in group one of the box classification.

1. Summerwood figured with wavy or branched radial bands. (Bands extend across the rings in the same direction as the rays.)
   (Compare Plate XVI, figures 1 to 3.)
   AA. Rays, many, broad, and conspicuous. They appear as "flecks" or "silver grain" on quarter-sawed material. Wood heavy to very heavy. Sapwood rather narrow. 40-49[1].   THE OAKS. 4[2]
   BB. Rays not noticeable. Color grayish brown, texture coarse. Sapwood narrow. Wood moderately light. 30. CHESTNUT. 1
2. Summerwood figured with short or wavy tangential lines (running more or less parallel with the rings), often most noticeable toward the outer part of the growth ring. (Compare Plate XVI, figures 4 to 8.)
   AA. Heartwood not distinctly darker than sapwood (sapwood sometimes darker than heartwood on account of sapstain.) Rays distinctly visible but fine. The wavy tangential lines conspicuous throughout the summerwood. Springwood pores numerous, in more than one row. Color pale to yellowish or greenish gray. Wood moderately heavy. **37.** HACKBERRY or SUGARBERRY. 4.
   BB. Heartwood distinctly darker than sapwood. Rays barely visible.
   (1) Springwood pores in **more than one row.**
      a. Very fine broken tangential lines visible in outer summerwood and especially prominent in wide rings. Sapwood several inches wide, heartwood brownish. Most pores or vessels except in outer sapwood appear somewhat closed, difficult to

---

[1]Figure indicates an average weight per cubic foot of the wood air dry, that is, containing 12 to 15 per cent moisture. U. S. Department of Agriculture Bulletin 556.

[2]This number indicates group to which the wood belongs in the box wood classification. Pages 100, 104.

blow through. Wood hard and heavy except pumpkin ash which usually is relatively soft, weak and brash. 36-44.

PUMPKIN ASH. `3·

WHITE ASH. 4.

GREEN ASH. 4.

b. Long and conspicuous wavy tangential bands throughout the summerwood. Sapwood very narrow. Heartwood brown with reddish tinge. Pores rather open. Wood moderately heavy. 37.

SLIPPERY ELM. 3.

(2) Springwood pores in **one** more or less continuous row except in wide rings where there are occasionally more. Heartwood brownish.

a. Pores in the springwood **fairly conspicuous** and visible, because of size and closeness together. Pores rather open. Wood moderately heavy. 36. WHITE ELM. 3.

b. Pores in the springwood **inconspicuous,** hardly distinguishable from those of the summerwood because relatively small, often not close together, and usually filled with tyloses. Wood heavy. 44.

CORK or ROCK ELM. 4.

3 Summerwood, generally **not** figured with radial or tangential bands. Rays barely visible. Several rows of large springwood pores which are usually open and easy to blow through. Sapwood narrow, rarely over three-fourths of an inch wide. Heartwood grayish to olive brown. Wood moderately heavy. 34. (Compare Plate XVI, figure 9.)

BLACK ASH. 3.

B. *Diffuse-porous woods*—No ring of large pores found at the beginning of each year's growth. Pores appear as fine grooves on the longitudinal cuts and are scattered with considerable uniformity throughout both the springwood and the summerwood. Pores vary in size from visible to the naked eye to barely visible or indistinguishable without a lens. The relatively small amount of difference in size between the springwood and summerwood pores makes it often difficult to distinguish the annual rings. Some of these woods are rather soft and light but are separated (because they contain pores or vessels) from "II," the conifers, or softwoods, which do not have true pores or vessels. Diffuse-porous woods are found in groups 1, 3, and 4 of the box woods. Those in group 1 are lightest. (Compare Plate XVI, figures 10 to 12, and Plate XVII, figures 1 to 11.)

AA. Individual pores plainly **visible.** Heartwood light chestnut brown. Sapwood narrow. Rays not visible on cross section. Wood light and soft. 27. (Compare Plate XVI, figure 10.)

BUTTERNUT. 1

BB. Individual pores barely visible. Sapwood wide. Rays not visible on cross section. (Compare Plate XV, figures 11 and 12.)

(1) Pores not crowded. Heartwood reddish brown. Wood moderately heavy to heavy. 38-44.

BIRCH. 4.

(2) Pores crowded. Heartwood grayish to brownish. Wood moderately light to light. 24-28.

COTTONWOOD. 1.
WILLOW   1.

CC. Individual pores **not** visible. (Compare Plate XVII, figures 1 to 11.)

(1) Rays comparatively broad and conspicuous, appear as flecks on quartered cuts and distinguish these woods from conifers. Color various shades of light reddish brown.

a. Rays crowded. No denser and darker band of summerwood noticeable. Wood usually lockgrained. Moderately heavy. 34.

SYCAMORE. 3

b. Rays not crowded. A distinct denser and darker band of summerwood present. Wood fairly straight-grained. Heavy. 44.

BEECH. 4.

(2) Rays not conspicuous but visible, hence distinguishing these woods from conifers.

a. Heartwood dingy reddish brown often with darker streaks. Sapwood pinkish white moderately wide, usually over an inch; often sold as "sap gum," sometimes stained blue by sapstain. Annual rings not clearly defined. Rays very fine, close together, not plain even on quartered cuts. Wood moderately heavy. 34.

RED GUM. 3.

b. Heartwood light reddish brown. Sapwood wide. Annual rings clearly defined by a thin darker reddish brown layer. Rays fine but distinct, conspicuous on quartered cuts because of darker color.

 (a) Wood hard, difficult to cut across the grain. Pith flecks rare. Rays appear to be not very close together as compared with soft maple. Wood heavy. 43.

 SUGAR OR HARD MAPLE. 4.

 (b) Wood comparatively easy to cut across grain. Pith flecks often abundant. Rays appear very close together as compared with hard maple. Wood relatively soft and only moderately heavy. 32-37.

 SOFT MAPLE. 3.

c. Heartwood pale to yellowish with a greenish, sometimes (especially in yellow poplar) purplish tinge. Sapwood usually over 1 inch wide. Annual rings clearly defined by a fine whitish line. Wood moderately light to moderately heavy. About 27-35.

 TULIP or YELLOW POPLAR. 1.
 CUCUMBER TREE. 1.
 MAGNOLIA. 1.

d. Heartwood pale or creamy brown often with scattered dark or black marks or streaks. Heartwood not sharply defined from light creamy colored sapwood. Wood light. 26.

 BASSWOOD. 1.

(3) Rays not distinctly visible on cross section. Annual rings usually not clearly defined which aids in distinguishing these woods from conifers.

a. Heartwood distinctly darker than sapwood.

 (a) Heartwood, reddish brown. Wood fairly straight-grained, pith flecks sometimes found. Pores visible in a good light, especially on longitudinal surfaces where they appear as fine lines or grooves. Wood moderately heavy to heavy. 38-44.

 BIRCH. 4.

(b) ˙Heartwood pale. to grayish brown. Wood often very cross grained. Moderately heavy. 34-35.

BLACK GUM. 3

TUPELO. 3

b. Heartwood not distinctly darker than sapwood. Wood odorless, tasteless.

(a) Color creamy. Annual rings inconspicuous, very faintly defined. Tangential surfaces show, when smoothly cut, faint fine bands running across the grain produced by the regularly spaced or storied rays "ripple marks." (See Plate XVII, figure 5.) Wood very light and soft. 25.

BUCKEYE. 1.

(b) Wood whitish. Annual rings clearly defined by a fine sometimes whitish line. **No** figure such as is produced by storied rays. Wood light. 28.

ASPEN or "POPPLE." 1.

Conifers

The softwoods, woods obtained from scale or needle-leaved trees. Woods without pores.

II. **No** pores present—Wood usually appears fine textured because the cells are small and regularly arranged and because no cells are strikingly larger than those surrounding them. **Annual rings are clearly defined by a definite band of summerwood.** Woods light, most of them in box wood group 1. A few heavier conifers make up group **2** of the box woods.

A. Odor and taste spicy-resinous. **No resin ducts, pitch pockets or accumulations of pitch present.**

THE CEDARS.

(Compare Plate XVIII, figures 1 and 2.)

1. Color creamy shading to a pale brown. Heartwood odor strong in green material, somewhat suggests ginger. Wood moderately light. 31.

PORT ORFORD CEDAR. 1.

2. Heartwood various shades of red and brown; odor resembling that of cedar shingles. Wood light to very light. **22.**   RED CEDAR. 1

ARBORVITAE. 1.

B. Odor and taste not spicy, may be resinous, especially in the pines. Pitch pockets and other accumulations of pitch, including small exudations on the ends of boards, often present. Knots usually more or less resinous. **Resin ducts present.**

1 Heartwood darker than sapwood.

AA. Resin ducts visible relatively conspicuous as small light specks on the cross section or as fine lines of slightly different color on the longitudinal surfaces. Wood with pitchy resinous odor or taste. Heartwood creamy to orange brown. (Compare Plate XVIII, figures 7 to 9.)

THE PINES.

(1) Summerwood relatively inconspicuous, not much harder or denser than springwood. Change from springwood to summerwood gradual. Heartwood pale creamy to light reddish brown. Resin ducts often conspicuous, especially in sugar pine. Wood moderately soft, light. 26-29.

WHITE PINE. 1.
SUGAR PINE. 1.

(2) Summerwood somewhat denser and more conspicuous than in (1). Color of heartwood reddish to orange brown. This group midway in density and appearance between (1) and (3). Weight 28-34.

WESTERN YELLOW PINE. 1.
LODGEPOLE PINE. 1.
JACK PINE. 1.
SCRUB PINE. 1.
NORWAY PINE. 1.

(3) Summerwood very dense, horny. Change from springwood to summerwood often very abrupt. Resin ducts to be seen, especially in or near the summerwood. Wood heavy. 35-45.

VIRGINIA AND CAROLINA PINE. 2.
SOUTHERN YELLOW PINE. 2.
PITCH PINE. 2.

BB.   Woods with rather inconspicuous resin **ducts,** without piny odor but with somewhat resinous odor and taste. Marked and rather abrupt change from springwood to summerwood. Pitch pockets or streaks may be found. (Compare Plate XVIII, figures 10 to 12.)

(1)   Color of heartwood usually reddish, sometimes with yellow cast. Summerwood dense. Scattered resin ducts present. Often several seen as small white dots in short tangential rows in or near the summerwood. Pitch pockets common. Wood moderately heavy. 30-34.

DOUGLAS FIR.   2·

Sometimes (DOUGLAS SPRUCE.
called   (OREGON PINE.

(2)   Heartwood dull russet brown. Summerwood sharply defined and fairly dense. Woods moderately heavy, especially that from butt cuts. 36.

LARCH.   2.
TAMARACK.   2·

(3)   Heartwood pale reddish. Transition from springwood to summerwood more gradual. Split tangential surfaces, especially if through the summerwood of narrow rings, characteristically indented or "dimpled." (See Plate XIX.) Split surfaces show "silky sheen." 26.

SITKA SPRUCE.   1

2.   Heartwood the same color as sapwood. Woods not conspicuously pitchy though resin ducts are present and pitch pockets may occur. Gradual transition from springwood to summerwood. (Split surfaces show "silky sheen.") Moderately heavy. 24-28

OTHER SPRUCES.   1.

C.   Wood without spicy odor, not pitchy or resinous. No **resin ducts,** pitch pockets or accumulations of resin normally present in the wood though resin may in some cases exude from the bark.

1.   Heartwood strongly colored. Summerwood dense.

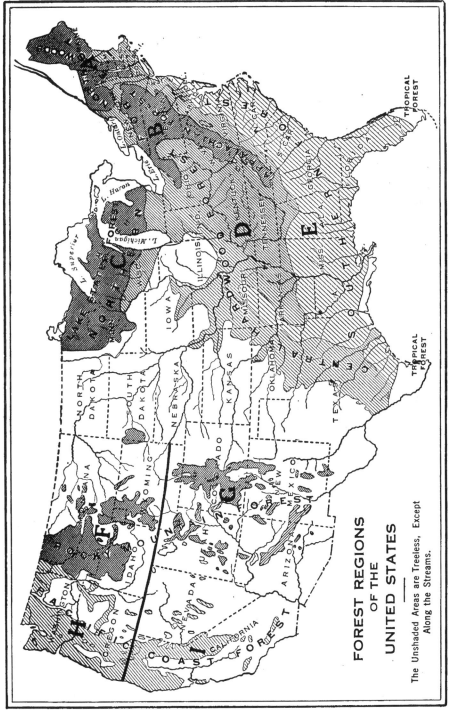

FOREST REGIONS
OF THE
UNITED STATES
—
The Unshaded Areas are Treeless, Except
Along the Streams.

FIG. 30—Forest regions of the United States.

    AA. Heartwood deep brownish red. Wood without markedly characteristic odor. Annual rings regular in width. Wood moderately light. 25-30.

REDWOOD. 1.

    BB. Heartwood light to very dark brown. Odor somewhat rancid. Longitudinal surfaces feel waxy. Annual rings very irregular in width. Weight variable. Average 30.

CYPRESS. 1.

  2. Heartwood not strongly colored.

    AA. Wood whitish at least in springwood. Summerwood darker, often sharply contrasted in color, tinged with red or purplish brown. Wood moderately light to light. 23-28.

THE TRUE FIRS. 1.

    BB. Wood has slight reddish hue in both springwood and summerwood. Wood splintery, often with cup shake. Odor somewhat sour when wood is fresh. Moderately light. 28.

HEMLOCK, 2.

DESCRIPTION OF BOX WOODS

The letters after the names refer to the various regions in which the trees grow as indicated on the accompanying map, figure 30, although the geographical distribution of each species is not confined exactly to the limits of the regions indicated. For scientific names see U. S. Dept. Ag. Bul. 17, "Check List of Forest Trees."

HARDWOODS

RING-POROUS WOODS

The Oaks—*White oak group* (A, B, C, D, E).

  Red oak group (A, B, C, D, E).

These species grow throughout the eastern half of the United States.

They are heavy and hard and when dry are without characteristic odor or taste. The annual rings are very distinct in both the white and the red oaks. Under a lens the pores of the summerwood of the white oaks are very minute and so numerous that they are difficult to count, but in the red oaks the opposite is true, which makes these species easy to distinguish. (Compare figures 1 and 2, Plate XVI.) The most characteristic feature of all oak woods is the presence of broad medullary rays very conspicuous on the end surface and appearing on the radial surface as silvery patches from ½ to 4 inches in height with the grain. This structure distinguishes the oaks from all other woods.

Chestnut (B, D).

The region of growth is the Appalachian highland and the central hardwood section.

The wood of chestnut is moderately light and usually straight-grained. The heartwood is grayish brown, with a slightly stringent taste due to the tannin in it. The annual rings are made very distinct by a broad band of porous springwood. The pores in the summerwood are very numerous and arranged in irregular radial bands similar to those in white oak. (See figure 3, Plate XVI.) The rays are much finer, however, than in the oaks.

The Elms—*White elm* (A, B, C, D, E)

Cork elm (B, D).

The range of white elm is the eastern half of the United States, that of cork elm being confined to the Appalachian highland and the central hardwood section.

The wood of white elm is moderately heavy and easy to work; that of cork elm is heavier, harder, and ranks higher in mechanical properties.

In both white and cork elm the springwood usually consists of but one row of large pores, those of the latter being smaller and filled with tyloses in the heartwood. (See figures 4 and 5, Plate XVI.)

Hackberry—(B, D, E, F, parts of G and H).

The range of growth includes Montana, Idaho, and the eastern half of the United States, except the northern portion.

The wood of hackberry is moderately heavy and generally straight-grained. The heartwood is light gray, tinged with green, which helps to distinguish this species from the elms in which the heartwood is brownish, usually with a reddish tinge. It is without characteristic odor or taste. The rays and annual rings are distinct without a lens which also helps to distinguish it from the elms. (See figure 7, Plate XVI.)

The Ashes—*White ash* (A, B, C, D, E).

Green ash (A, B, C, D, E).

Black ash (A, C, and northern part of B and D).

Pumpkin ash (Parts of E and western D).

The range of white and green ash is the eastern half of the United States; of black ash, the northern part of this section, and that of pumpkin ash, the southern portion. White and green ash are very much alike and are sold as "white ash" or "ash." The lighter weight grades of white and green ash are commercially classed as "pumpkin ash."

The sapwood of each is comparatively wide and white. The heartwood is grayish brown occasionally with a reddish tinge. In black ash the sapwood is narrow, usually less than one inch wide and the heartwood is silvery or olive brown, resembling that of chestnut. Black ash averages considerably lighter in weight than the other two species.

· All species have definite annual rings made very conspicuous by several rows of large pores in the springwood. In the summerwood the pores are few, very small, and isolated, or occasionally two or three in a radial row. Except in black ash, these pores are surrounded by light colored tissue which projects tangentially, producing light-colored lines often joining pores somewhat separated, especially in the outer portion of the annual rings. (See figures 8 and 9, Plate XVI.) Elm can be distinguished from ash by the arrangement of its numerous summerwood pores in wavy tangential lines.

DIFFUSE-POROUS WOODS

Butternut—(A, B. D).

The range of growth is the northeastern part of the United States, including the central hardwood section.

Butternut resembles black walnut in structure but is lighter in weight, softer and lighter colored, resembling black ash or chestnut in this respect. It differs from the woods previously discussed in that it has no very pronounced group of springwood pores.

Birch—(A, B, C, D, E).

These species grow chiefly in the northern portion of the United States, east of the Mississippi River.

The structure of the different birches is very similar. The wood is heavy, fairly straight-grained and without characteristic odor or taste. The pores are barely visible to the naked eye on the cross section but quite readily visible as grooves on the longitudinal surfaces. The annual rings, because of the almost uniform size of the pores, are rather indistinct. Pith flecks are very often present in birches.

The birches may be confused with the maples. In the

maples, however, the rays on the cross section are visible to the naked eye, while in the birches they are not. Furthermore, the pores of the maples are not visible to the naked eye, while those of the birches can generally be seen when the wood is examined in a good light.

Cottonwood—(B, C, D, E, F).

Aspen—(A, B, C, D, F, G, H, I).

The range of growth of cottonwood is all of the eastern section of the United States except New England and the northern part of the Rocky Mountain section; that of aspen is all sections of the United States except the South Atlantic and Gulf States.

Both species are light, fairly straight-grained, and without characteristic odor or taste.

There is practically no difference in color between the sapwood and heartwood of either species. The pores are larger in cottonwood than in aspen. The rays are not readily visible to the naked eye. (See figure 12, Plate XVI, and figure 3, Plate XVII.)

Cotton gum or tupelo resembles cottonwood but usually is heavier and has smaller pores. Yellow poplar is similar in weight and hardness but its greenish tinge usually distinguishes it. Basswood has more of a creamy white color, smaller pores, and distinct rays.

Sycamore—(A, B, C, D, E).

The range is the eastern half of the United States.

The wood is moderately heavy, usually lock-grained, without characteristic odor or taste. The heartwood is colored from light to a moderately dark reddish brown, sometimes not clearly defined from the sapwood. The pores are very small and crowded together. The rays are very characteristic; they are comparatively broad and conspicuous, although not as large as the largest rays in the oaks. They are all practically of the same size. On the radial surface they appear as reddish brown "flakes," similar to the rays in oak, but smaller.

Sycamore is not easily confused with other woods. Its conspicuous rays and interlocked grain make it easily recognizable. (See figure 6, Plate XVII.) It resembles beech somewhat, but can be distinguished from it by the rays, only a small portion of which are broad in beech. Beech also is heavier and has a distinct dense band of summerwood.

Beech—(A, B, D, E, eastern half of C).

This species grows in the eastern section of the United States and also in northern Wisconsin

Beech is a hard heavy wood without characteristic odor or taste. The heartwood has a reddish tinge varying from light to moderately dark. The pores are invisible without a lens and decrease in size slightly and gradually, from the inner to the outer portion of each ring. (See figure 7, Plate XVII.) Some of the rays are broad, being fully twice as wide as the largest pores and appearing on the radial surface as reddish brown flakes. The other rays are very fine. The maple resembles beech, except that in maple the widest rays are about the same width as the largest pores and not so conspicuous on the radial surface.

Red gum—(D, E, and part of B).

Red gum grows in the Appalachian section and in the Gulf States.

It is moderately heavy, somewhat lock-grained, and without characteristic odor or taste. The sapwood is white with a pinkish hue or often blued with sapstain. The heartwood is reddish brown, often with irregular darker streaks. The wood has a very uniform structure. The annual rings are inconspicuous and pores are not distinct to the unaided eye, but the rays are fairly distinct without a lens. (See figure 8, Plate XVII.) The uniform structure, interlocked grain, and reddish brown color are usually sufficient to distinguish red gum from other woods.

The Maples—*Sugar maple* (A, B, C, D, E)

The range of growth of this species is the eastern half of the United States.

Sugar maple is heavy, hard and difficult to cut across the grain, in which respect it differs from the softer maples. The sapwood is white in all maples, and the heartwood is light reddish brown, without characteristic odor or taste. The annual rings are defined by a thin reddish layer usually more conspicuous on dressed longitudinal surfaces. The pores are all very small and uniformly distributed throughout the annual ring. The rays are distinct without a lens and on radial surfaces they are conspicuous as small reddish brown flakes. In sugar maple only part of the rays are as wide as the pores; the others are very fine, being barely visible with a lens. The differences between the soft and hard maples are similar to those distinguishing sycamore and beech, although in the maples the rays are finer. In soft maple the rays are crowded as in sycamore. Birch and beech resemble maple somewhat. Birch has larger pores, visible as fine grooves on the dressed surfaces and the rays on the end surfaces are not distinctly visible without a lens. In beech some of the rays are very conspicuous.

Yellow poplar—(B, D, E).

The range of growth is the Appalachian highland, the central hardwood section and Gulf States.

Yellow poplar is moderately light, straight-grained, and without characteristic odor or taste. The heartwood is light to a moderately dark yellowish or olive brown with a greenish and sometimes purplish tinge or streaks. The annual rings are outlined by light-colored lines. The pores are evenly distributed throughout the annual ring, and are too small to be visible to the unaided eye. The rays are distinct without a lens. (See figure 1, Plate XVII.)

Cucumber tree—(B, D, E).

The cucumber tree grows in the same locality as yellow poplar except in Florida and the South Atlantic Coast.

It is easily confused with yellow poplar and is usually sold as such, although it averages slightly heavier in weight. It is much inclined to stain.

Basswood—(A, B, C, D).

This species grows in the eastern half of the United States.

It is a light, soft, straight-grained wood with a creamy brown color. The heartwood is not clearly defined from sapwood. Sometimes black or brownish spots or streaks are present. It is without taste, but has a slight characteristic odor even when dry. The pores are invisible without a lens. The rays are fairly distinct on the end surface. (See figure 11, Plate XVII.)

Cottonwood resembles basswood, but is more grayish in color, has larger pores, and very fine rays. Buckeye also resembles basswood in color and texture, except that the rays are much finer and are visible on the cross section only with a good lens. They form characteristic so-called "ripple marks" on the tangential surfaces. (See figure 4, Plate XVII.)

The Gums—*Black gum* (A, B, D, E).

Cotton gum (southern D, E).

Black gum grows in the eastern half of the United States,

except in northern Michigan, Wisconsin, and Minnesota, and cotton gum in the Southern Atlantic and Gulf States.

These woods are moderately heavy to heavy, very lock-grained. They are without characteristic odor or taste. The annual rings are indistinct. The rays and pores are not distinct to the naked eye. The weight, lock grain, and lack of well defined summerwood distinguish these woods from the conifers. Their lack of distinctive characters assists in their identification. Cotton gum is often lighter and softer in the butt log than near the top. Cottonwood sometimes resembles tupelo, or cotton gum, but its more visible pores serve to identify it. (See figure 2, Plate XVII.)

Yellow buckeye—(D).

The range of growth of this species is the Appalachian highland, the Ohio valley, and into Texas.

This wood is light, soft, straight-grained, and without characteristic odor or taste. The heartwood is not clearly differentiated from the sapwood. The general color of the wood is creamy white or yellowish. The annual rings are often not clearly defined. The pores and rays are not visible to the naked eye, although characteristic "ripple marks," produced by groups of rays, may be seen on tangential surfaces. (See figures 4 and 5, Plate XVII.) Buckeye resembles basswood but the rays in basswood, though fine, can be distinguished more readily than those in buckeye. Buckeye also somewhat resembles aspen.

CONIFERS (NON-POROUS WOODS)

The Cedars—*Western red cedar* (H)

In the United States this species grows in the northwestern part, chiefly in Washington and Oregon.

The western red cedar is light and straight-grained. The heartwood is reddish-brown, with the characteristic odor of cedar shingles and a somewhat bitter taste when chewed. The wood contains no resin ducts, although it contains a small quantity of aromatic oils. The annual rings are distinct, moderate in width, with a thin, but well defined band of summerwood. Pores are entirely absent, and the rays are very fine. (See figure 2, Plate XVIII.)

Northern white cedar (A. B. C).

The range of growth is the northern part of the United States from Maine to Minnesota.

This species resembles western red cedar in odor and

taste, but usually it is without the reddish hue, has very narrow annual rings, and averages lighter in weight.

Port Orford cedar (H).

This species grows in southwestern Oregon and northwestern California.

It is a moderately light, straight-grained wood with a pronounced odor and taste; the odor is sometimes compared to that of ginger. The wood is less spongy than that of some of the other cedars. The odor and light color make the identification of this wood easy. (See figure 1, Plate XVIII.)

The White Pines—*Eastern white pine* (A, B, C and parts of D).

Western white pine (F, H, I).

The region of growth of eastern white pine in the United States is the northern part from Maine to Minnesota, and that of western white pine, the northwestern part.

The wood of both these species is moderately light, straight-grained and practically tasteless, but has a slight, yet pleasant and distinct resinous odor. The heartwood is creamy to light reddish or yellowish brown. The annual rings are distinct, but the summerwood is not a pronouncedly darker or appreciably harder layer. The outer portion of western yellow pine logs often has narrow annual rings with a very thin layer of summerwood so that this species may approximate the white pines in appearance, and consequently is often sold as white pine. It may be distinguished, however, by its horny glistening layers of summerwood, which are especially prominent in the wider rings.

Sugar pine (I)

Sugar pine grows in the northern part of California and in Oregon.

It is very much like the white pines in structure and properties, and in fact belongs to the white pine group botanically. The heartwood is very light brown, only slightly darker than the sapwood and practically never reddish, as is the case, quite often, in the white pines. The summerwood never appears as a horny, glistening band as in the hard pines. The wood of sugar pine has a slightly coarser texture than that of white pine; that is, the fibers and also the resin ducts have a greater average diameter. Resinous exudations, which become granular and have a sweetish taste, are quite common in rough sugar pine lumber, and when present are the more reliable means of distinguishing it from the other pines. (See figure 7, Plate XVIII.)

The Yellow or Hard Pine Group—*Western yellow pine* (F, G H, I).

Norway pine (A, C, northern half of B)

Southern yellow pine (E).

These species range as follows: western yellow pine, in the Rocky Mountain and Pacific slopes; the southern yellow pine, in the South Atlantic and Gulf States, and Norway pine in the Northeastern and Central States.

The yellow pines are mostly heavier, harder, more resinous, and contain a wider and harder layer of summerwood than the white pines. However, exceptions occur, notably western yellow pine, which in the outer part of the mature trees is often as light in weight as the average white pine. In the different species and even in the same species the sapwood is variable in width, averaging narrowest in some species of southern yellow pine. The heartwood is orange-brown to reddish-brown color. The summerwood is usually defined as a conspicuously denser, harder, and darker band, but in very narrow rings such as are found in the sapwood of old trees of western yellow pine the summerwood layer may be very narrow and inconspicuous. The resin ducts are visible with a lens on a smoothly-cut end surface, and may be seen as brownish or whitish lines on the longitudinal surfaces. Douglas fir is somewhat similar to yellow pine in appearance, but usually has a distinct reddish hue and less prominent resin ducts.

Douglas fir—(F, G, H, I)

Douglas fir grows abundantly in the Pacific Northwest and throughout the Rocky Mountain region.

It differs from the true firs in being more resinous, heavier, stronger and in having a distinctly darker heartwood. The annual rings are made distinct by a conspicuous band of summerwood. Resin ducts are present, but not so distinct as in the pines, usually appearing on a cross section as whitish specks in the summerwood. (See figure 10, Plate XVIII.)

The Spruces—*White spruce* (A, B, C).

Red spruce (A, B).

Sitka spruce (H).

Engelmann spruce (G, F).

White spruce grows in the northern part of the United States east of the Mississippi, red spruce in the same section, except in the western part, Sitka spruce in western Washington and Oregon, and Engelmann spruce in the Rocky Mountain highland.

These species are moderately light, straight-grained woods. In the white, red, and Engelmann spruce, the heartwood is as light colored as the sapwood, but in Sitka spruce the heartwood has a light reddish tinge, making it a little darker than the sapwood. The annual rings are clearly defined by a distinct but not horny band of summerwood. Spruce resembles the white pines in texture but has a silky sheen. (See Plate XIX.) On account of its reddish tinge Sitka spruce might be confused with light grades of Douglas fir from which it can be distinguished, however by the pocked or dimpled appearance of split tangential sur faces. (See Plate XIX.) Douglas fir has denser summer wood except in very narrow rings; therefore, rings of aver age width should be compared. Engelmann spruce is somewhat lighter in weight and weaker than the other spruces.

Bald cypress (E).

This species grows abundantly in the South Atlantic and Gulf States.

It is highly variable in color and weight. Commercially, the common cypress is classed as "white," "yellow," "red," or "black" cypress, although it is all derived from the same botanical species. The wood has a characteristic rancid odor when fresh. In dry wood the odor is less pronounced, but can be detected by sawing it and holding the sawdust to the nostrils. The wood is without characteristic taste. The annual rings usually are irregular in width and outline. The summerwood is very distinct but narrow, although wider than in the cedars. (See figure 3, Plate XVIII.) Cypress resembles the cedars and redwood somewhat; but the cedars have an aromatic odor and spicy taste, and redwood is tasteless and odorless.

Redwood (I, along the coast)

Redwood grows in the coast region of northern California.

It is moderately light, straight-grained and obtainable in large clear pieces. The heartwood, as a rule, is deep reddish brown in color. Occasionally, lighter colored pieces, resembling western cedar, are found. The wood contains no resin ducts. The annual rings are made very distinct by dense bands of summerwood alternating with soft, spongy springwood. (See figure 6, Plate XVIII.) Redwood is without characteristic odor or taste.

The True Firs—*Balsam fir* (A, B, C).
Noble fir (H).
White fir (G, I)
Red fir (I).
Alpine fir (F, G, H).

With the exception of balsam fir these species grow abundantly in the Rocky Mountain highland and on the Pacific slopes.

They are all moderately light, straight-grained and with the exception of Alpine fir, practically without characteristic odor or taste. Alpine fir has, when dry, a distinctly disagreeable odor. The color of the wood is whitish, often with a reddish brown tinge, which is especially noticeable in the summer wood bands. This produces a sharp color contrast in each ring which is a very distinctive character in most of the woods of this group. The wood is very uniform. Rarely short lines of resin ducts resulting from injury may be found. The firs resemble hemlock but the weight and difference in color between springwood and summerwood are often sufficient to distinguish them.

Hemlock—*Hemlock* (A, B, C, D)
Western hemlock (F, H).

Hemlock grows in the eastern half of the United States, except in the southeast portion. The range of growth of western hemlock is the northwestern part of the United States.

These woods are about medium in weight but are grouped with the heavier conifers for box and crate construction. They are usually straight-grained, sometimes twisted, and the eastern species is often splintery and subject to cup shakes. When fresh, hemlock has a characteristic sour odor, but this practically disappears when the wood is dry. There is little difference in color between sapwood and heartwood, although the latter may have a somewhat pale brown or reddish hue. There is no striking contrast in color between the springwood and summerwood such as is generally found in the firs, the change of color in the hemlocks is gradual. The rays in hemlock are not visible to the naked eye and the wood does not normally have resin ducts. The western hemlock is less splintery and subject to cup shakes than the eastern species. Tangential lines of abnormal resin ducts caused by injuries are present, especially in the western species.

GRADING RULES FOR ROTARY-CUT BOX LUMBER

The following rules, revised May 20, 1919, are used by the Rotary Cut Box Lumber Association:

Specifications shall always be furnished by buyer to manufacturer as follows:

Thickness ....First
Width across grain............Second
Length with grain..............Third

1.   All stock shall be log run, the full cut of the log, and shall be free from rot or dote. Pin-worm holes, sound tight knots, discoloration, and stain are no defect.

2·   All stock shall be machine-cut to thickness, standard gears as furnished by lathe manufacturers to be used.

3.   All stock shall be cut tight, and, when shipped, shall weigh not to exceed 3,100 pounds per thousand board feet if kiln-dried, or 3,400 pounds per thousand board feet if air dried, railroad weights at point of shipment to govern. Stock shall be sufficiently flat to straighten under machine without splitting

· 4.   A trimming allowance of ½ inch in length shall be made on all stock up to 30 inches long and of 1 inch on stock longer than 30 inches, all lengths to have ½-inch trimming allowance in width; but if not to exceed 25 per cent in any one car shall measure scant of the ½-inch trimming allowance in widths, but full ¼ inch, it shall be considered up to specifications.

5.   All cut-downs in width that accumulate in cutting out defects and rounding logs shall be accepted by buyer, these cut-downs to run in 2-inch multiples down to 4 inches, unless otherwise agreed; but not over 25 per cent of contents of any car, feetage basis, shall consist of these cut-downs. When sawed after drying, these cut-downs may be exact width; but if they are sized green, a ½-inch trimming allowance, when dry, shall be made.

· 6.   Checks or splits not longer than one-fourth the length of the piece, but in not more than 15 per cent of the pieces in each shipment, are allowed, provided these checks or splits are reasonably straight, or do not diverge more than 2 inches per foot, and do not run over ½ inch in width on pieces 18 inches and up wide, not over ⅜ inch on pieces 12 to 18 inches wide, not over ¼ inch on pieces 6 to 12 inches wide, and not over ⅛ inch on pieces 6 inches and under wide.

Splits or checks ⅛ inch and under wide are not considered defects.

7. Specifications on all sizes, both width and length, shall not be divided in fractions of less than ¼ inch.

Other Species—The minor kinds of boxwoods are graded as follows: The hardwoods, sycamore, ash, etc., may be ordered as No. 2 common, according to the rules of the two hardwood associations. Larch covers eastern tamarack (Northern Pine Manufacturers' Association rules) and western larch (Western Pine Manufacturers' Association rules). Noble fir, white fir, and red fir are admitted in No. 3 common under the West Coast Lumbermen's Association rules. In California, white fir and red fir are sold with the box grades of California pine. Cedar includes southern red (no standard rules); northern white (no standard rules), southern white (no commercial rules but see Navy specification 3903b), and western red cedar (West Coast Lumbermen's Association and Pacific Lumber Inspection Bureau). Redwood may be ordered as merchantable in accordance with the rules of the California Redwood Association.

APPENDIX

Table 15. Cement-coated Coolers or Standard Nails and Sinkers or Countersunk Nails

| Size | Number of nails per keg | Length inches | Gauge of wire | Dimension of Heads | | | | Net weight per keg pounds |
| | | | | Coolers[2] | | Sinkers[2] | | |
				Diameter inches	Thickness inches	Diameter inches	Thickness inches	Pounds
2d	85,700	1	16	11/64	.016	5/32		79[4]
3d	54,300	1⅛	15½	3/16	.013	3/16		64[4]
4d	29,800	1⅜	14	7/32	.029	13/64		61[4]
5d	25,500	1⅝	13½	15/64	.023	7/32		70[4]
6d	17,900	1⅞	13	1/4	.027	7/32		65[4]
7d	15,300	2⅛	12½	1/4	.025	15/64		72[4]
8d	10,100	2⅜	11½	19/64	.025	17/64	Sinkers have slightly countersunk heads—impossible to give thickness	71
9d	8,900	2⅝	11½	9/32	.029	17/64		68
10d	6,600	2⅞	11	5/16	.033	9/32		63
12d	6,200	3⅛	10	11/32	.030	21/64		80
16d	4,900	3¼	9			21/64		83
20d	3,100	3¾	7	See footnote 5	See footnote 5	11/32		84
30d	2,400	4¼	6			13/32		82 to 85
40d	1,800	4⅝ and 4¾	5			15/32		
50d	1,300	5¼	4			1/2		79
60d	1,100	5⅝ and 5¾	3			9/16		82

[1]Coolers and sinkers are identical in dimensions and count, differing only in the heads.

[2]The cooler head is flat and of good size, preferred by many for machine-driving and in work where large diameter is desired. It is perfectly satisfactory for hand-driving in soft wood.

[3]The sinker head is reinforced by a slight countersinking, which reduces the diameter. It will not break or pull off, and is recommended for hand-driving in hardwood. It can be used in automatic nailing machines.

[4]Weights of some manufactures are one-half pound more.

[5]The sinker type of head is used on coolers larger than 12d; the larger sizes are, therefore, identical in every particular.

Either of these types may be used for box and crate work if regular cement-coated box nails are not available.

TABLE 16. CEMENT-COATED BOX NAILS[1]

Size	Number of nails per keg	Length inches	Gauge of wire	Diameter at head	Thickness of head	Net Weight per keg
				Inches	Inches	Pounds
2d	96,200	⅞ to 1	16½	5/32	.016	67½ to 74
3d	64,600	1⅛	15½ or 16	3/16	.016	68 to 80
4d	45,500	1⅜	15½	3/16	.017	64 to 72¼
5d	39,700	1⅝	15	7/32	.016	74 to 78
6d	23,600	1⅞	13½ to 14	1/4	.022	67½ to 77
7d	19,300	2⅛	13 to 13½	1/4	.022	69 to 72
8d	14,000	2¼ to 2⅜	12½	17/64	.024	70 to 75
9d	13,100	2½ to 2⅝	12½	17/64	.034	71 to 78
10d	8,900	2⅞	11½	9/32	.037	69½ to 75
12d	8,700	3⅛	11½	9/32	.031	80 to 80½
16d	7,100	3⅜	11	5/16	.030	78
20d	5,200	3⅞	10	11/32	.036	83
30d	4,600	4⅜	10	3/8	.030	81
40d	3,500	4⅞	9	7/16	.034	84

[1]The variation in some values is due to the difference in the manufacturing specifications of the several manufacturers.

TABLE 17. MISCELLANEOUS CEMENT-COATED NAILS

Type of nail	Size	Number of nails per keg	Length inches	Gauge of wire	Size of head		Net weight per keg pounds
					Diameter inches	Thickness inches	
Egg Case......	2d	73,500	1	16	3/16	.013	70
"	3d	51,700	1⅛	15	1/4	.020	70
"	4d	30,500	1½	14	9/32	.024	70
Orange Box....	4d	55,000	1¼	15	13/64	.020	81
Fruit Box......	4d	45,500	1⅜	15	13/64	.011	73
Apple Box.....	5d	31,000	1⅝	14	7/32	.019	74
Berry Box.....	¾"	140,000	¾	17	5/32	.011	70
" "	⅞"	125,000	⅞	17	5/32	.013	70
Veneer Box[1] ...	4d	30,500	1½	14	9/32	.018	70
Veneer[2]	4d	19,700	1⅜	12	5/16	.020	70
....	5d	13,000	1⅝	11	3/8	.021	70
....	6d	9,000	1⅞	10	1/2	.029	70
Barrel.........	¾"	95,000	¾	15½	....	....	70
....	⅞"	67,200	⅞	14½	....	....	70
....	1	55,000	1	14½	....	....	70
....	1⅛	48,000	1⅛	14½	....	....	70
....	1¼"	37,000	1¼	14	....	....	70
"	1⅜"	26,000	1⅜	13	....	....	70
"	1½	24,500	1½	13	....	....	70

[1]For hoopless orange boxes.
[2]For use in 3-ply veneer packing cases.

NAMES AND DESCRIPTION OF GRADES OF LUMBER SUITABLE FOR PACKING BOXES

White pine[1]

ITEM	White Pine Association of the Tonawandas, North Tonawanda, N. Y.	Northern Pine Manufacturers Association, Minneapolis, Minn.	Western Pine Manufacturers Association, Spokane, Wash.
Box grade	No. 1 box	No. 4 common	Western (Idaho) white pine. No. 4 common
	This grade admits coarse knots, regardless of size and not necessarily sound, also a reasonable amount of shake or stain.	1. The predominating defect characterizing this grade is red rot. 2. Other types are pieces showing numerous large worm holes, or several knot holes, or pieces that are extremely coarse, knotted, waney, shaky, or badly split. 3. Pieces which are extremely cross checked are admissible in this grade.	1. The predominating defects characterizing this grade are red rot and knot holes. 2. Other types are pieces showing numerous large worm holes, pieces that are extremely coarse, knotted, waney, or showing excessive heart shake, extremely pitchy, or badly checked or split.
Higher grade	No. 3 barn	No. 3 common	No. 3 common

[1] No standard rules for New England white pine box lumber are in effect at the present time (1921).

White pine—Continued

ITEM	White Pine Association of the Tonawandas, North Tonawanda, N. Y.	Northern Pine Manufacturers Association, Minneapolis, Minn.	Western Pine Manufacturers Association, Spokane, Wash.
Lower grade	No. 2 box	No. 5 common	No. 5 common
		Short box—includes lumber 12 to 47 inches long inclusive, 3 inches and wider, and No. 4 and better.	
Thicknesses, inches:			
Rough	1, 1¼, 1½, 2, 2½, 3, 4.		
S1S⎫ S2S⎭	¹⁵⁄₁₆⎰ Not adopted by ⅞⎱ Ass'n.	⎰1, 1¼, 1½, 2. ²⁵⁄₃₂, 1⅛, 1⅜, 1¾. ⎱Same as S1S.	⎰1, 1¼, 1½, 2. ²⁵⁄₃₂, 1⅛, 1⅜, 1¾. ⎱Same as S1S.
Widths, inches.	4, 5, 6, 8, 10, 12, 13 and wider, 4 to 16 with average of 9 or better, 4 to 12 averaging at least 8 may be specified.	Mixed widths, 4 and wider, or in specified widths of 4, 6, 8, 10, 12, 13 and wider. Average of 9 may be specified if ordered 4 to 16. Best to order 4 to 12 averaging at least 8. S2E—½ off.	Sold in mixed widths, 4 and wider. S2E—½ off.
Length, feet.	6, 8, 10, 12, 14 and 16. 6 to 16 averaging at least 12 may be specified.	6, 8, 10, 12, 14, 16, 18 or 20. 6 to 16 averaging at least 12 may be specified.	Sold in mixed lengths, 6 and longer. 6 to 16 averaging at least 12 may be specified.

Yellow pine, including North Carolina pine

ITEM	Southern Pine Association, New Orleans, La. Georgia-Florida, Sawmill Association, Jacksonville, Fla.	North Carolina Pine Association, Norfolk, Va.
	Southern yellow pine.	North Carolina pine.
Box grade	No. 2 common	Box
	No. 2 common boards dressed one or two sides; admits knots not necessarily sound, but the mean or average diameter of any one knot must not be more than one-third of the cross-section if located on the edge and must not be more than one-half of the cross-section if located away from the edge; a sound knot may extend over one-half the cross-section if located on the edge, except that no knot the mean or average diameter of which exceeds 4 inches is admitted; admits also worm holes, splits one-fourth the length of the piece, wane 2 inches wide, or through heart shakes one-half the length of the piece, through rotten streaks ½ inch wide one-fourth the length of the piece, or its equivalent of unsound red heart; or defects equivalent to the above. A knot hole 3 inches in diameter will be admitted, provided the piece is otherwise as good as No. 1 common. Miscut 1-inch common boards which do not fall below ¾ inch in thickness are admitted in No. 2 common, provided the grade of such thin stock is otherwise as good as No. 1 common.	Lumber below the grade of No. 3, containing pinholes, pin, standard, and large, reasonably sound knots, stain not exceeding 25 per cent, and pith knots, encased knots, and spike knots which do not seriously affect strength of piece; stained pieces otherwise No. 1 and 2 grade, which show over 50 per cent stain, and stained pieces otherwise grading No. 3 and showing not more than 33⅓ per cent stain; and pitchy pieces which are not desirable in No. 1, 2 and 3 grades. Lumber which would otherwise grade No. 1, 2 and 3 containing 50 per cent firm red heart will be admitted in this grade Reverse side of box boards may be cull.

Yellow pine—Continued

ITEM	Southern Pine Association, New Orleans, La. Georgia-Florida, Sawmill Association, Jacksonville, Fla.	North Carolina Pine Association, Norfolk, Va.
Higher grade	No. 1 common	No. 3
Lower grade	No. 3 common	Culls and merchantable red heart.
Thicknesses, inches:		
Rough	1, 1¼, 1½.	1, 1¼, 1½, 1¾, 2.
S1S	¹³⁄₁₆, 1¹⁄₁₆, 1⁵⁄₁₆.	⅞, 1⅛, 1¼, 1½, 1¾
S2S	Same as S1S.	¹³⁄₁₆, 1¹⁄₁₆, 1¼, 1½, 1¾.
Widths, inches:	3 and up, in multiples of 1, not over ½ inch scant on 8 and under, ⅝ on 9 or 10 and ¾ on 11 and 12 or wider.	Stocks—6, 8, 10 and 12, Edge —random widths under 12 except 6, 8, 10 inches, which are stocks. 4/4 edge is 3 and wider, 5/4 to 8/4 edge is 4 and wider. ¼ in width shall be allowed for dressing 6 and under boards four sides, but ½ shall be allowed for, dressing boards wider than 6.
Lengths, feet:	S. P. A. 4 to 24 in multiples of 2. G.—F. S. A. 8 to 20 in multiples of 1.	8 to 16 in multiples of 2, not exceeding 5 per cent of 8 foot.
	An average of 15 may be specified.	An average of 12 may be specified.

Spruce

ITEM	West Coast Lumbermen's Association, Seattle, Wash. Pacific Coast Lumber Inspection Bureau, Inc.	Spruce Manufacturers' Association (Not active although grading rules are still used).	
Box grade	Sitka spruce box lumber. The value and grade of this lumber is determined by its adaptability for the manufacture of ordinary packing boxes, ordinary sizes being defined as boxes not over 20 inches in length nor more than 15 inches in width. Wide boards or those of special widths will admit more defects than narrow or random widths. Grades—There are three recognized grades of box lumber, viz.: No. 1, No. 2 and No. 3. No. 1—Generally sound, and contains from 75 per cent to 90 per cent of cuttings suitable for boxes of ordinary size and quality, as referred to above. In computing percentages, cuttings of assorted sizes are used. Assorted sizes are defined as pieces running in widths from 6 inches to 12 inches, and in lengths from 12 inches to 20 inches.	Appalachian spruce box. Large black knots, knots not sound in character, knot holes, heart checks or shakes, black sap and small amount of hard red wood admitted. Wane or bark equal to half the thickness and one-fourth the length on the face or equal to 20 per cent of the piece on the back, admitted. Season checks or splits equal to ⅓ the length of the piece admitted. Pin worms and scattering grub holes admitted. This grade is designed for boxes and crating and some waste or bad material is allowed.	New England spruce. No standard grading rules are in effect. The lumber is graded principally according to verbal understanding between buyer and seller or according to local specifications which have been in use for a number of years. Massachusetts State Law for the inspection of Lumber. Box boards, waney edged box boards, pine, bass wood, poplar and spruce are inspected as good and culls. Good includes all sound lumber so free from black, mouldy, or rotten sap, rot, worm holes and bad shakes, that not less than ⅔ of the entire piece (as a whole) can be used without waste. Culls include all lumber not good enough for the above grade. The Navy Department has the following rule for New England spruce (39S1b.):

Spruce—Continued

ITEM	West Coast Lumbermen's Association, Seattle, Wash. Pacific Coast Lumber Inspection Bureau. Inc.	Spruce Manufacturers' Association(Not active although grading rules are still used).	
	Sitka spruce box lumber—Cont'd. No. 2—Generally similar in character to No. 1, containing 60 per cent to 75 per cent of box cutting. No. 3—All lumber below the grade of No. 2 and containing 40 per cent to 60 per cent of box cuttings.		New England spruce—Cont'd. Box: (a) Sizes:— Lengths to be random 6 feet and up. Widths to be 4 inches and up as specified. Thicknesses as specified. (b) Defects allowed —This grade will admit the following: Large branch and black knots, knot holes, worm holes, stained sap, and a reasonable amount of hard red rot. Wane not exceeding one-half the thickness of piece and extending full length on one edge only or a proportional amount on two edges. Shakes, splits or checks equal to ⅓ length of piece.
Higher grade	No. 1 common.	Merchantable.	
Lower grade	None.	Mill culls.	
Thicknesses, inches: Rough	1, 1¼, 1½, 2.	1, 1¼, 1½, 2.	
S1S or S2S	¾, 1¹⁄₁₆, 1⁵⁄₁₆, 1¾	¹³⁄₁₆, 1⅛, 1⅜, 1¾.	
Widths, inches: Rough	4, 6, 8, 10, 12.	4 and wider.	
S2E	3½, 5½, 7½, 9½, 11½.	3⅜, 5⅝, 7⅝, 9½, 11½.	
Length, feet	6 to 20 in multiples of 2.	6 and longer. Not over 5 per cent 6 feet.	

Red and sap gum

ITEM	National Hardwood Lumber Association, Chicago, Ill. American Hardwood Manufacturers' Association, Memphis, Tenn.
Box grades	No. 2 common red and sap gum. Pieces 3 to 7 inches wide, 4 to 10 feet long must work 50 per cent clear face or sound sap in not over three cuttings; pieces 3 to 7 inches wide, 11 feet and longer must work 50 per cent clear red face or sound sap in not over four cuttings; pieces 8 inches and over wide, 4 to 9 feet long must work 50 per cent clear red face or sound sap in not over three cuttings; pieces 8 inches and over wide, 10 to 13 feet long must work 50 per cent clear red face or sound sap in not over four cuttings; pieces 8 inches and over wide, 14 feet and over long must work 50 per cent clear red face of sound sap in not over five cuttings. No cutting to be considered which is less than 3 inches wide by 2 feet long. Sound discolored sap is no defect in any grade of sap gum.
Higher grade	No. 1 common.
Lower grade	No. 3 common.
Thicknesses	See Table 9.
Widths, inches	3 and over in random widths, average of 9 may be specified.
Lengths, feet	4 and over, but not more than 10 per cent 4 and 5 foot lengths admitted in this grade. Average of 11 may be specified.

Western yellow pine

ITEM	Western Pine Manufacturers' Association, Spokane, Wash.	California White and Sugar Pine Manufacturers' Association.
	"Western white pine."	"California white pine."
Box grade	No. 4 common.	No. 3 common and fencing.
	1. The defects common to this grade are much the same as those in No. 3 but greater. 2. The most common serious defects are knot holes, and either red rot, or its equivalent in heavy massed pitch. Other types are: extremely coarse knotted, or waney, or badly split, or badly checked pieces, or pieces with excessive heart shake. 3. This grade especially meets the demands of the box manufacturer for a soft, easily-worked pine in a grade that yields well in cut-up box product.	The general appearance of this grade of lumber is coarse. It admits large, loose or unsound knots, an occasional knot hole, some shake, worm holes, some red rot, any amount of stained sap, but not a serious combination of these defects in any one piece. A grade similar to No. 3 common is sometimes sold as No. 1 Box in California. No. 4 common and strips. The predominating defects characterizing this grade are red rot, pitch, and stain. Other types are pieces showing numerous large worm holes, or pieces that are extremely coarse knotted, waney, shaky, or badly split. A grade similar to No. 4 common is sometimes sold as No. 2 box.
Higher grade	No. 3 common.	No. 3 common.
Lower grade	No. 5 common.	None.
Thicknesses, inches: Rough S1S or S2S	1, 1¼, 1½, 2. $\frac{25}{32}$, 1⅛, 1⅜, 1¾.	1, 1¼, 1½, 2. ⅞, 1$\frac{5}{32}$, 1$\frac{13}{32}$, 1$\frac{25}{32}$.
Widths, inches	4 and wider. If S2E, ¼ scant.	4 and wider.
Length, feet	6 and longer.	6 and longer.

Cottonwood

ITEM	National Hardwood Lumber Association American Hardwood Manufacturers' Association
Box grade	No. 2 common.
	Pieces 3 to 7 inches wide, 4 to 10 feet long must work 50 per cent sound in not over three cuttings; pieces 3 to 7 inches wide, 11 feet and longer must work 50 per cent sound in not over four cuttings; pieces 8 inches and over wide, 4 to 9 feet long must work 50 per cent sound in not over three cuttings; pieces 8 inches and over wide, 10 to 13 feet long must work 50 per cent sound in not over four cuttings; pieces 8 inches and over wide, 14 feet and over long must work 50 per cent sound in not over five cuttings. No cutting to be considered which is less than 3 inches wide by 2 feet long.
Higher grade	No. 1 common.
Lower grade	No. 3 common.
Thicknesses	See Table 9.
Widths, inches Lengths, feet	3 and over, average of 9 may be specified. 4 and over, not to exceed 10 per cent of 4 and 5 foot lengths. Average of 11 may be specified.

Yellow poplar

ITEM	National Hardwood Lumber Association American Hardwood Manufacturers' Association
Box grades	No. 2-B common.
	No. 2 common is divided into No. 2-A common and No. 2-B common, but unless otherwise specified is to be considered as a combined grade.
	Sound discolored sap is no defect in this grade.
	No. 2-B common—pieces 3 to 7 inches wide, 4 to 10 feet long must work 50 per cent sound in not over three cuttings; pieces 3 to 7 inches wide, 11 feet and longer must work 50 per cent sound in not over four cuttings; pieces 8 inches and over wide, 4 to 9 feet long must work 50 per cent sound in not over three cuttings; pieces 8 inches and over wide, 10 to 13 feet long must work 50 per cent sound in not over four cuttings; pieces 8 inches and over wide, 14 feet and over long must work 50 per cent sound in not over five cuttings. No cutting to be considered which is less than 3 inches wide by 2 feet long.
Higher grade	No. 2-A common and No. 1 common.
Lower grade	No. 3 common.
Thicknesses	See Table 9.
Width, inches	3 and over, average of 9 may be specified.
Lengths, feet	4 and over, not more than 10 per cent of 4- and 5-foot lengths. Average of 9 may be specified.

Hemlock

ITEM	West Coast Lumbermen's Association	Northern Hemlock and Hardwood Manufacturers' Association	Eastern States Hemlock
Box grade	No. 2 common.	Box and crating inch and dimension.	
	Must be free from rot. Admits large, coarse knots approximately 2 inches in diameter in 4-inch and 6-inch stock, 2½ inches in 8 and 10-inch, and ⅓ the width of the piece in 12-inch and wider, spike knots, any amount of solid heart or sap stain, a limited number of well scattered worm holes, solid pitch or pitch pockets, small amount of fine shake, wane 2 inches wide, if it does not extend into the opposite face. A serious combination of above defects in any one piece is not permitted. A board may have one large knot hole, provided the piece is otherwise as good as No. 1 common.	Stock that will cut at least 50 per cent of firm, useful box and crating stock. Includes No. 3 hemlock, 4/4 and 8/4, 2 inches and wider, 4 feet and longer, and admits defects of the following character: Soft rot, open shake, coarse loose knots and knot holes, or any other defect that is characteristic of hemlock, that will weaken stock to the extent of barring its use for dimension purposes. This grade must be based on the percentage of useful material that each piece contains, as it is impossible to describe the defects which this stock contains.	Spruce Manufacturers' Association rules sometimes used in West Virginia and North Carolina. In Pennsylvania, New York, and New England local rules for "Box" followed.

Hemlock—Continued ·

ITEM	West Coast Lumbermen's Association	Northern Hemlock and Hardwood Manufacturers' Association	Eastern States Hemlock
Higher grade	No. 1 common.	Select No. 3 common.	
Lower grade	No. 3 common.	No. 4 common.	
Thicknesses, inches:			
Rough	1, 1¼, 1½, 2.	$\frac{15}{16}$, 1⅞.	
S1S	¾, 1$\frac{1}{16}$, 1$\frac{5}{16}$, 1¾.	$\frac{25}{32}$, 1¾.	
S2S	Same as S1S.	$\frac{25}{32}$, 1⅝.	
Widths, inches	4, 6, 8, 10, 12.	2, 4, 6, 8, 10, 12.	
S1E	3½, 5½, 7¼, 9¼, 11¼	¼, to ⅜ scant.	
S2E	Same as S1E.	⅜ scant.	
Lengths, feet.	10 to 20.	4 to 20 in multiples of 2.	

Soft maple and soft elm

ITEM	National Hardwood Lumber Association American Hardwood Manufacturers' Association
Box grade	No. 2 common.
	Pieces 3 to 7 inches wide, 4 to 10 feet long must work 50 per cent sound in not over three cuttings; pieces 3 to 7 inches wide, 11 feet and over long must work 50 per cent sound in not over four cuttings; pieces 8 inches and over wide, 4 to 9 feet long must work 50 per cent sound in not over three cuttings; pieces 8 inches and over wide, 10 to 13 feet long must work 50 per cent sound in not over four cuttings; pieces 8 inches and over wide, 14 feet and over long must work 50 per cent sound in not over five cuttings.
	No cutting to be considered which is less than 3 inches wide by 2 feet long.
Higher grade	No. 1 common.
Lower grade	No. 3 common.
Thicknesses	See Table 9.
Widths, inches	3 and over, average of 7 may be specified.
Lengths, feet	4 and over, not over 10 per cent 4- and 5-foot lengths. Average of 11 may be specified.

Birch

ITEM	National Hardwood Lumber Association American Hardwood Manufacturers' Association
Box grades	No. 2 common. Pieces 3 to 7 inches wide, 4 to 10 feet long must work 50 per cent clear face in not over three cuttings; pieces 3 to 7 inches wide, 11 feet and over long must work 50 per cent clear face in not over four cuttings; pieces 8 inches and over wide, 4 to 9 feet long must work 50 per cent clear face in not over three cuttings; pieces 8 inches and over wide, 10 to 13 feet long must work 50 per cent clear face in not over four cuttings; pieces 8 inches and over wide, 14 feet and over long must work 50 per cent clear face in not over five cuttings. No cutting to be considered which is less than 3 inches wide by 2 feet long.
Higher grade	No. 1 common.
Lower grade	No. 3 common.
Thicknesses	See Table 9.
Widths, inches	3 and over, average of 7 may be specified.
Lengths, feet	4 and over, not over 10 per cent 4- and 5-foot lengths. Average of 11 may be specified.

Basswood

ITEM	National Hardwood Lumber Association American Hardwood Manufacturers' Association
Box grades	No. 2 common. Pieces 3 to 7 inches wide, 4 to 10 feet long must work 50 per cent sound in not over three cuttings; pieces 3 to 7 inches wide, 11 feet and over long must work 50 per cent sound in not over four cuttings; pieces 6 inches and over wide, 4 to 9 feet long must work 50 per cent sound in not over three cuttings; pieces 8 inches and over wide, 10 to 13 feet long must work 50 per cent sound in not over four cuttings; pieces 8 inches and over wide, 14 feet and over long must work 50 per cent sound in not over five cuttings. No cutting to be considered which is less than 3 inches wide by 2 feet long.
Higher grade	No. 1 common.
Lower grade	No. 3 common.
Thicknesses	See Table 9
Widths, inches	3 and over. Average of 9 may be specified.
Lengths, feet	4 and over, not over 10 per cent 4- and 5-foot lengths. Average of 11 may be specified.

Beech

ITEM	National Hardwood Lumber Association American Hardwood Manufacturers' Association
Box grades	No. 2 common. Pieces 3 to 7 inches wide, 4 to 10 feet long must work 50 per cent clear face in not over three cuttings; pieces 3 to 7 inches wide, 11 feet and over long must work 50 per cent clear face in not over four cuttings; pieces 8 inches and over wide, 4 to 9 feet long must work 50 per cent clear face in not over three cuttings; pieces 8 inches and over wide, 10 to 13 feet long must work 50 per cent clear face in not over four cuttings; pieces 8 inches and over wide, 14 feet and over long must work 50 per cent clear face in not over five cuttings. No cutting to be considered which is less than 3 inches wide by 2 feet long. Wormy beech. Shall be graded according to the rule for beech No. 2 common and better, with the exception that pin worm holes shall not be considered a defect.
Higher grade	No. 1 common.
Lower grade	No. 3 common.
Thicknesses	See Table 9.
Widths, inches	3 and over. Average of 7 may be specified.
Lengths, feet	4 and over, not over 10 per cent 4- and 5-foot lengths. Average of 11 may be specified.

Tupelo and black gum

ITEM	National Hardwood Lumber Association American Hardwood Manufacturers' Association Southern Cypress Manufacturers' Association, New Orleans, La.
Box grades	No. 2 common.
	There is no restriction as to heart in No. 2 common tupelo. Sound discolored sap is no defect. Pieces 3 to 7 inches wide, 4 to 10 feet long must work 50 per cent sound in not over three cuttings; pieces 3 to 7 inches wide, 11 feet and longer must work 50 per cent sound in not over four cuttings; pieces 8 inches and over wide, 4 to 9 feet long must work 50 per cent sound in not over three cuttings; pieces 8 inches and over wide, 10 to 13 feet long must work 50 per cent sound in not over four cuttings; pieces 8 inches and over wide, 14 feet and over long must work 50 per cent sound in not over five cuttings. No cutting to be considered which is less than 3 inches wide by 2 feet long.
Higher grade	No. 1 common.
Lower grade	No. 3 common.
Thicknesses	See Table 9.
Widths, inches	3 and over. Average of 7 may be specified.
Lengths, feet	4 and over, not over 10 per cent 4- and 5-foot lengths. Average of 10 may be specified.

Oak (plain, red, or white)

ITEM	National Hardwood Lumber Association American Hardwood Manufacturers' Association
Box grade	No. 2 common. Pieces 3 to 7 inches wide, 4 to 10 feet long must work 50 per cent clear face in not over three cuttings; pieces 3 to 7 inches wide, 11 feet and longer must work 50 per cent clear face in not over four cuttings; pieces 8 inches and over wide, 4 to 9 feet long must work 50 per cent clear face in not over three cuttings; pieces 8 inches and over wide, 10 to 13 feet long must work 50 per cent clear face in not over four cuttings; pieces 8 inches and over wide, 14 feet and over long must work 50 per cent clear face in not over five cuttings. No cutting to be considered which is less than 3 inches wide by 2 feet long.
Higher grade	No. 1 common.
Lower grade	No. 3 common
Thicknesses	See Table 9.
Widths, inches	3 and over. Average of 7 may be specified.
Lengths, feet	4 and over, not to exceed 10 per cent of 4- and 5-foot lengths. Average of 10 may be specified.

Balsam fir

ITEM	Northern Pine Manufacturers' Association	New England balsam fir
	May be purchased with white pine.	Usually sold with spruce.

Cypress

ITEM	National Hardwood Lumber Association	Southern Cypress Manufacturers' Association American Hardwood Manufacturers' Association
Box grades	No. 1 boxing.	Box.
	Must work 66⅔ per cent in cuttings containing not less than 72 square inches. No cutting considered which is less than 18 inches long or less than 3 inches wide. Each cutting may contain sound stain, pin worm holes, unsound knots and peck that do not extend through the piece, season checks, and other defects that do not prevent the use of the cutting for boxing purposes.	Each piece must contain 66⅔ per cent or more of sound cuttings, no single cutting to contain less than 72 square inches. No piece of cutting may be shorter than 2 feet or narrower than 3 inches. Sound cuttings will admit all the defects allowed in No. 1 common The waste material may be thin or absolutely worthless.
	No. 2 boxing.	
	This grade may contain all lumber not admitted in No. 1 boxing, but each piece must work at least 50 per cent in the same size cuttings described in No. 1 boxing.	
Higher grade	No. 2 common.	No. 2 common.
Lower grade	Peck.	Peck.
Thicknesses, inches:		
Rough	Same as for hardwoods.	1, 1¼, 1½, 2.
S1S	See Table 9.	¹³⁄₁₆, 1⅛, 1⅜, 1¾.
S2S		Same as S1S.
Widths, inches	Random widths 3 and over. Average of 7 may be specified.	Random widths 3 and wider. Average of 7 may be specified S1E, ⅜ off; S2E ½ off.
Length, feet	No. 1 boxing, 6 and over. No. 2 boxing, 4 and over. Equal proportions may be specified.	6 to 20. Average of 12 may be specified.

Chestnut

ITEM	National Hardwood Lumber Association American Hardwood Manufacturers' Association
Box grades	No. 2 common. Pieces 3 to 7 inches wide, 4 to 10 feet long must work 50 per cent sound in not over three cuttings; pieces 3. to 7 inches wide, 11 feet and over long must work 50 per cent sound in not over four cuttings; pieces 8 inches and over wide, 4 to 9 feet long must work 50 per cent sound in not over three cuttings; pieces 8 inches and over wide, 10 to 13 feet long must work 50 per cent sound in not over four cuttings; pieces 8 inches and over wide, 14 feet and over long must work 50 per cent sound in not over five cuttings. No cutting to be considered which is less than 3 inches wide by 2 feet long. Sound Wormy Worm holes admitted in this grade without limit. No piece shall contain heart to exceed ¾ its length in the aggregate. Pieces 4 inches wide, 6 and 7 feet long must be sound; 8 to 11 feet long must work 66⅔ per cent sound in not over two pieces; 12 feet and over long must work 66⅔ per cent sound in not over three pieces. No piece of cutting to be less than 2 feet long by the full width of the piece. Pieces 5 inches and over wide, 6 to 11 feet long must work 66⅔ per cent sound in not over two pieces; 12 feet and over long must work 66⅔ per cent sound in not over three pieces. No piece of cutting considered which is less than 4 inches wide by 2 feet long or 3 inches wide by 3 feet long.
Higher grade	No. 1 common.
Lower grade	No. 3 common.
Thicknesses	See Table 9.
Widths, inches	No. 2—3 and over. Sound wormy, 4 and over. Average of 8 may be specified.
Lengths, feet	No. 2—4 and over, not to exceed 10 per cent of 4- and 5-foot lengths. Sound wormy, 6 and over, not to exceed 10 per cent of 6- and 7-foot lengths. Average of 11 may be specified.

Sugar pine

Same as western yellow pine (California white pine) on page 148 under rules of California White and Sugar Pine Association.

Plate I—Defects recognized in the commercial grading of

PLATE I

Pin knots   Standard knot   Large knot   Encased knot

Pith knot   Rotten knot   Spike knot   "Heart" Pith

Shake   Checks   Pitch pocket Upper: Edge-grain Pitch pocket Lower: Flat-grain   Pitch streak

Stained sap   Water stain   Mineral streaks   Gum spots

Bird pecks   Rot   Shot-worm holes Grub-worm holes   Rafting-pin hole Knot hole

Wane   Loosened grain   Insufficient thickness Chipped grain   Method of measuring knots "A" is the effective diameter

PLATE II

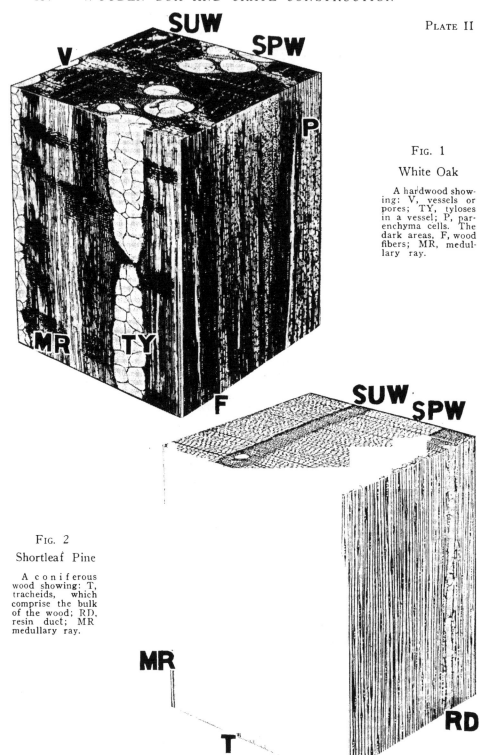

FIG. 1

White Oak

A hardwood showing: V, vessels or pores; TY, tyloses in a vessel; P, parenchyma cells. The dark areas, F, wood fibers; MR, medullary ray.

FIG. 2

Shortleaf Pine

A coniferous wood showing: T, tracheids, which comprise the bulk of the wood; RD, resin duct; MR medullary ray.

II—Cubes of wood magnified about 25 diameters.

G. 1—White oak.

G. 2—Shortleaf pine.

In each cube the top view represents the transverse or end surface, the left view the radial or "quartered" surface, and the right view the tangential or plain-sawed surface. (SPW) springwood; (SUW) summerwood.

The medullary rays are continuous from the starting point to the bark, and the vessels are continuous longitudinally, although the illustrations show them interrupted.

Plate III—Styles of wooden boxes, nailed and lock-corner

PLATE III

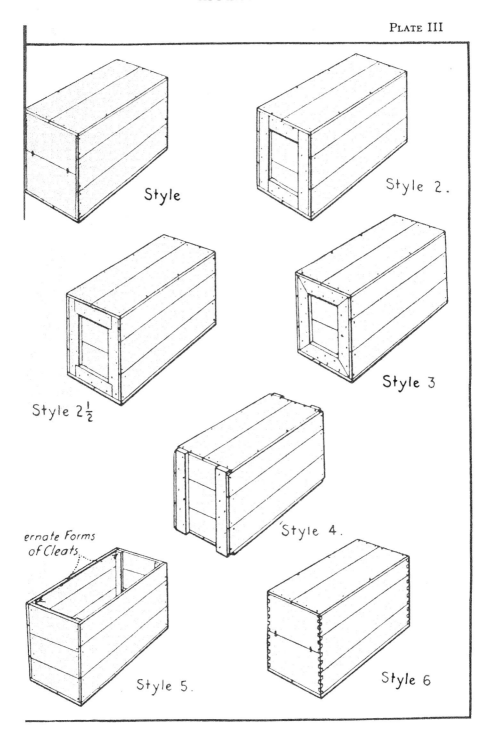

Style

Style 2.

Style 2½

Style 3

Style 4.

ernate Forms
of Cleats

Style 5.

Style 6

Plate IV—Special styles of boxes.

Figs. 1 and 3—Boxes made for easy opening.
Fig. 2—An accessible box (cover screwed on and sealed with
Fig. 4—Modified Style 2 box adapted to be closed without na
Fig. 5—Modified Style 4 box adapted to be closed without na

PLATE IV

FIG. 1   FIG. 2

FIG. 3

FIG. 4   FIG. 5

PLATE V—Strapped boxes.

Fig. 1—Reinforced battens.

Fig. 2—End-opening box, cover held only by strapping.

Fig. 3—Double corner nails to keep the straps in position occurs.

Figs. 4 and 5—Common methods of box strapping.

PLATE V

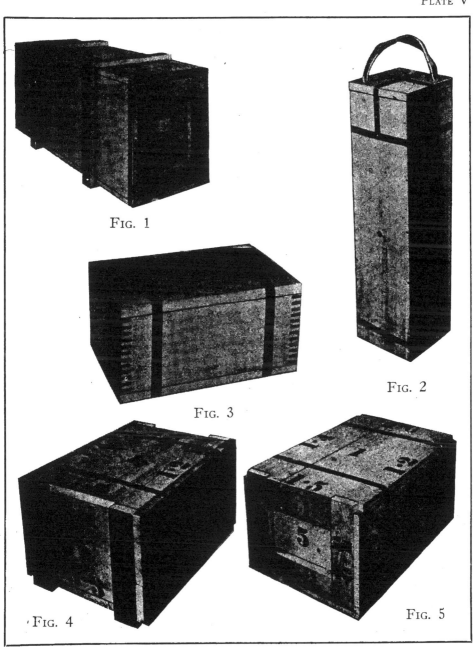

Fig. 1

Fig. 2

Fig. 3

Fig. 4

Fig. 5

PLATE VI—Types of handles.

FIG. 1—Handhold for boxes in which the opening is not objectionable.

FIG. 2—Section of a box end with handhold formed by beveling the edge of cleat.

FIG. 3—Handhold for boxes of medium weight that need to be handled with care.

FIG. 4—Method of attaching rope handles.
Right—Outside end of box.
Left—Reverse side of cleats.

FIG. 5—Method of attaching webbing handles.
Upper—Outside face of end.
Lower—Inside face of end.

FIG. 6—Another method of attaching webbing handles.
Left—Outside face of end.
Right—Inside face of end.

Plate VI

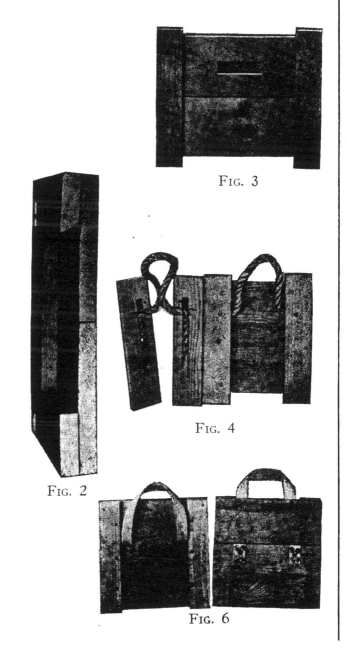

Fig.

Fig. 3

Fig. 2

Fig. 4

Fig. 5

Fig. 6

PLATE VII—Different types of corner construction.
  FIG. 1—Details of dovetail corner.
  FIG. 2—Dovetail box corner.
  FIG. 3—Test specimens for determining holding power of with and at an angle to grain.
  FIG. 4—Joints for 4-one box cleats.
    . Upper—Mortise and tenon.
    Lower—Step mitre.

PLATE VII

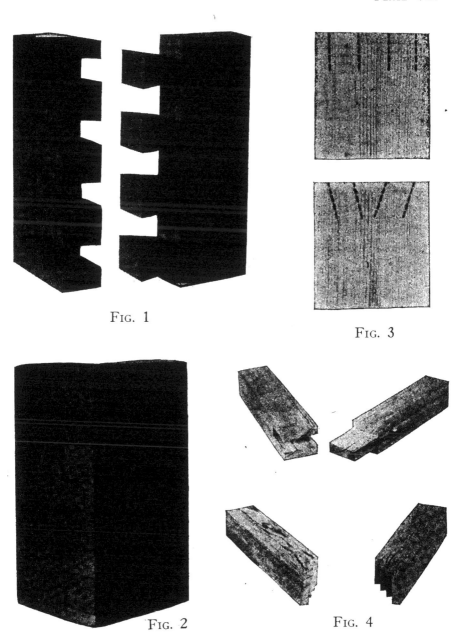

FIG. 1

FIG. 3

FIG. 2

FIG. 4

PLATE VIII—Wirebound boxes.

FIG. 1—Fassnacht box—note method of joining the wire corners.

FIG. 2—Method of reinforcing battens. (c) Regular cleats with and tenoned joints, (b) Battens, (n) Cement-coated (

FIG. 3—"4-one" wirebound box closed for shipment.   Note and character of twists for uniting the binding wires.

FIG. 4—The outer surface of a mat for a "4-one" box.

FIG. 5—"4-one" box with inside liners or corner cleats.

FIG. 6—"4-one" wirebound box assembled.

PLATE VIII

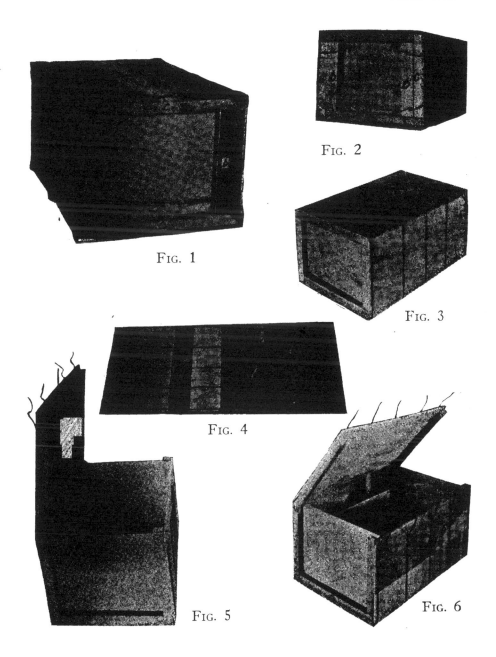

FIG. 2

FIG. 1

FIG. 3

FIG. 4

FIG. 5

FIG. 6

Plate IX—Types of commercial boxes.

    Fig. 1—Phonograph box made of plywood fastened to a frame.

    Fig. 2—Egg crate with sides, top, and bottom made of rotary-cut veneer.

    Fig. 3—Apple box.

    Fig. 4—Orange case.

    Fig. 5—Upright piano box.

PLATE IX

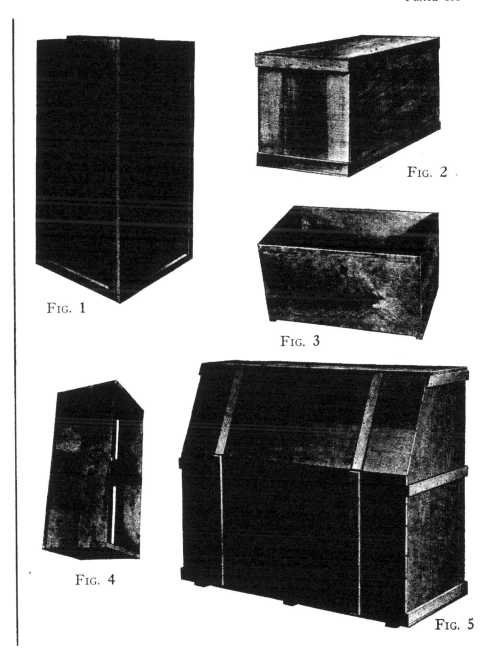

FIG. 1

FIG. 2

FIG. 3

FIG. 4

FIG. 5

PLATE X—Types of plywood or veneer panel boxes.

FIG. 1—This box corresponds in some points of construct Style 3 of nailed box.

FIG. 2—Corresponds in some respects with hardware type shown in Plate XIV.

FIG. 3—Corresponds most closely to Style 2 of nailed box.

FIG. 4—This box has double number of cleats of box shown but is otherwise similar in construction.

Plate X

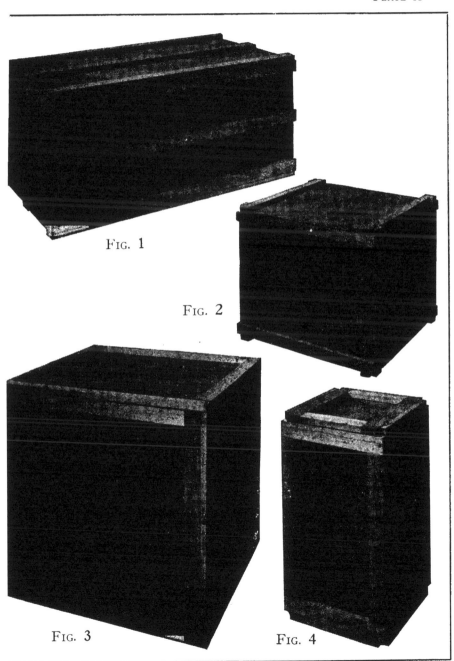

Fig. 1

Fig. 2

Fig. 3

Fig. 4

Plate XI—Three-way crate corners.

Fig. 1—Three-way crate corner made of 2⅝ by 3¾ inch stock.

Fig. 2—Three-way crate corner made of 2 by 6 inch stock.

Fig. 3—Three-way crate corner made of 2⅝ by 3¾ inch stock.

Fig. 4—Method of cutting and nailing a diagonal brace.

Fig. 5—A common style of crate corner.

Fig. 6—A style of crate corner consisting of the corner as shown in Fig. 5 with one double member.

Fig. 7—One method of reinforcing a crate corner

PLATE XI

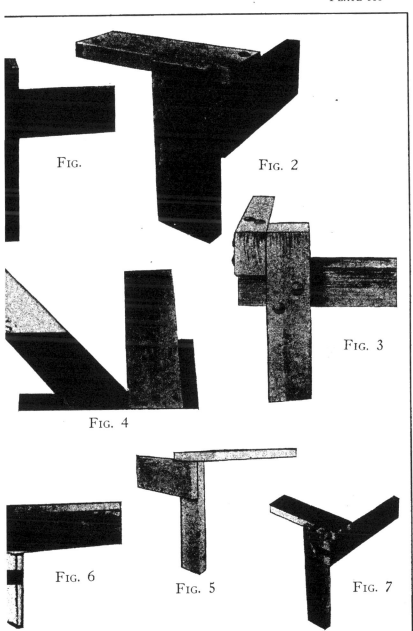

Fig.   Fig. 2

                         Fig. 3

Fig. 4

Fig. 6   Fig. 5   Fig. 7

PLATE XII—Various arrangements of crate members at three-wa

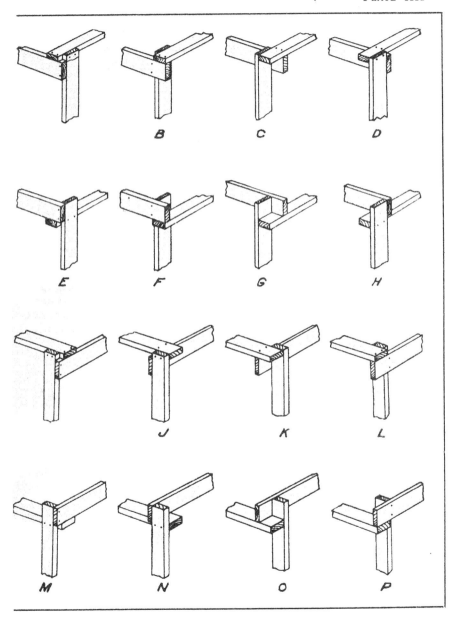

B

C

D

E

F

G

H

J

K

L

M

N

O

P

PLATE XIII—Crates with special features.

FIG. 1—Several sets of cross-bracing and cross-members used crease rigidity and bending strength.

FIG. 2—The scabbing shown inside the vertical members extends outer edges of the top and bottom horizontal crate members

FIG. 3—Framework of crate to which all other parts are fa

FIG. 4—Crate with 3-way corner construction showing cross-b diagonal bracing, and extra pieces to strengthen the skids support the vertical members.

PLATE XIII

FIG. 1

FIG. 2

FIG. 3

FIG. 4

PLATE XIV—Method of numbering faces of test boxes and crates f
    convenience in recording data and location of failures.

FIGS. 1 and 2—Crate numbering system.
FIGS. 3 and 4—Box numbering system.

PLATE XIV

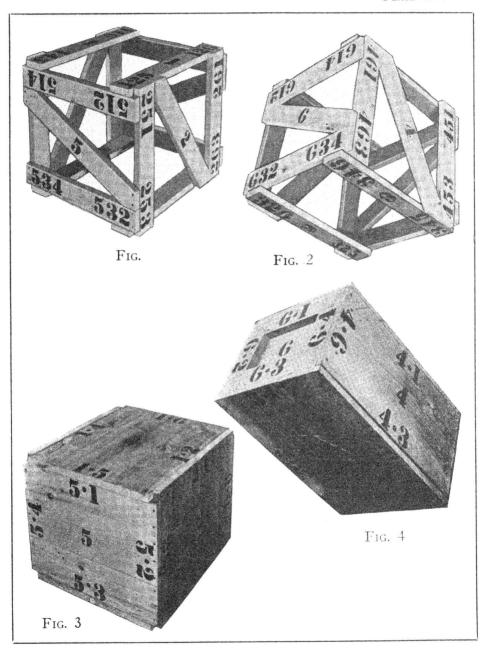

FIG.

FIG. 2

FIG. 4

FIG. 3

PLATE XV—Different kinds of joints and **fasteners.**

FIG. 1—Use of corrugated fasteners.
FIG. 2—Different types of corrugated fasteners.
FIG. 3—Dovetail joint.
FIG. 4—Coil of corrugated fastening material.
FIG. 5—Lock-corner joint.
FIG. 6—Lap joint.

PLATE XV

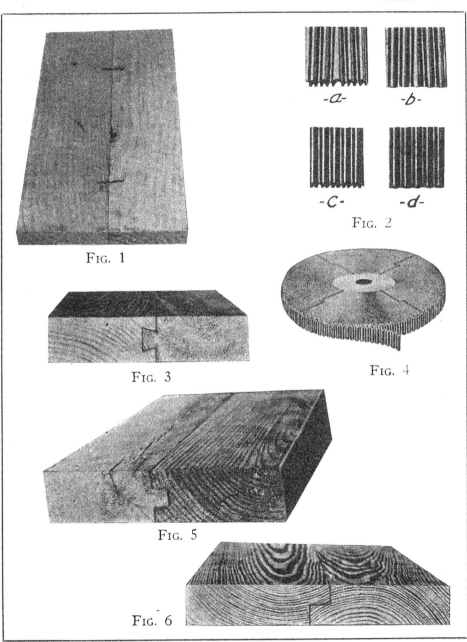

FIG. 1

-a- -b-

-c- -d-

FIG. 2

FIG. 3

FIG. 4

FIG. 5

FIG. 6

PLATE XVI—Hardwoods with pores.

Ring-porous hardwoods, figures 1-9 inclusive,
    Summerwood with radial figure, figures 1-3.
    Summerwood with tangential figure, figures 4-8.
    Summerwood without special figure, figure 9.

Diffuse-porous hardwoods, figures 10-12 inclusive,
    Pores easily visible to naked eye, figure 10.
    Pores barely visible to naked eye, figures 11-12.

PLATE XVI

FIG. 2—A white oak.

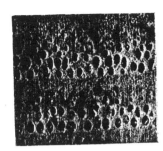

FIG. 3—Chestnut.

FIG. 5—Cork or rock elm.

FIG. 6—Slippery elm.

FIG. 8—A white ash.

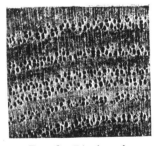

FIG. 9—Black ash.

FIG. 11—Birch.

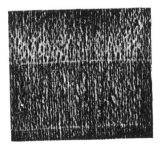

FIG. 12—Cottonwood.

PLATE XVII

Fig. 1—Yellow poplar.   Fig. 2—Black gum.   Fig. 3—Aspen, or "pop-ple."

Fig. 4—Buckeye.   Fig. 5—Ripple marks on tangential surface as in buckeye.

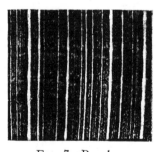

Fig. 6—Sycamore.   Fig. 7—Beech.   Fig. 8—Red gum.

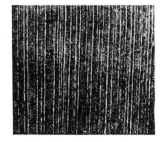

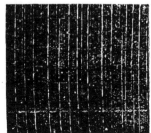

Fig. 9—Soft maple.   Fig. 10—Hard maple   Fig. 11—Basswood.

-porous hardwoods.

s.e woods are not readily visible to the naked eye.
y in size:
in figures 6 and 7
us but visible, figures 1, 8, 9, 10 and 11.
visible figures 2, 3, 4.

PLATE XVIII—Softwoods, conifers, or woods without pore

Woods without resin ducts, figures 1-6.
Woods with resin ducts, figures 7-12.

PLATE XVIII

FIG. 2—Western red cedar.

FIG. 3—Cypress.

FIG. 5—Hemlock.

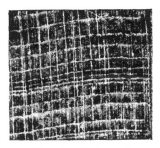

FIG. 6—Redwood.

FIG. 8—Western yellow pine.

FIG. 9—Southern yellow pine.

FIG. 11—Larch.

FIG. 12—Spruce.

Fig. 1—Sitka spruce

ential surfaces f Sitka spruce and Douglas fir.

"dimpled" appearance of the spruce, not found in
This characteristic is most pronounced in Sitka
arrow rings, and is almost entirely absent in very
aterial

INDEX

A

Air-drying of ˌlumber. As decay
　　preventive, 32, 36
　time necessary for, 36
Allowance for shrinkage, 17
Alpine fir. Identification, 136
Annealed strapping, 60
Annual rings, 110
Appalachian spruce. Rules for
　　grading, 145
Arborvitae. Structure, 122
Army ordnance boxes, 43
Ash. Identification, 127
　structure, 119
　see also Black ash; Green ash;
　　Pumpkin ash; White ash
Aspen, *see* Popple
Assembling of wirebound boxes, 105
　with detached tops, 107
　with wedgelock ends, 107
Association grading rules, 11
Availability of box lumber, 1, 41

B

Balanced construction, 43
　how determined, 87
　relation to nailing qualities of
　　wood, 51, 53
　spacing of nails in, 55
Bald cypress. Identification, 135
Balsam fir. Identification, 136
　rules for grading, 159
Barbed nails, 55
Basswood. Identification, 131
　rules for grading, 155
　structure, 121
Battens. In crates, 81
　in wirebound boxes, 68, 107
　use in strapping, 61
Beech. Identification, 130
　rules for grading, 156
　structure, 120
Binding rods in crates, 84
Binding wire for wirebound boxes,
　105
Birch. Identification, 128
　rules for grading, 154
　structure, 121
Bird pecks, 10

Black ash. Identification, 127
　structure, 119
Black gum. Identification, 131
　rules for grading, 157
　structure, 122
Blemishes in lumber, 8
Blue stain, 31, 110
Bolting qualities of wood, 81
Bolts. Carriage, 83
　machine, 84
　use in crates, 83
Bored holes for nails. Effect, 83
　for lag screws, 84
Bow, 10
Box design, 40, balance in 43
　characteristic of various styles,
　　64
　defined, 40
　factors determining size, 73
　factors determining strength re-
　　quired, 70
　factors influencing details, 40
　limitations by traffic rules, 74, 85
　of boxes with detached tops, 107
　of hinged boxes, 62
　of wirebound boxes, 103
　one-piece parts, 44
　relation to available equipment,
　　42
　relation to strapping, 61
　relation to wood groups, 103
　special constructions, 74
　with wedgelock ends, 105
Boxes. Salvage value, 1
　second hand, 2
Box nails, 53, 101
Box styles, 42
Box lumber, *see* Lumber; Woods.
Braces. Fitting and fastening, 80
　internal, 84
　use on long crates, 79
Buckeye. Structure, 122
　see also Yellow buckeye
Butt joint, 44
Butternut. Identification, 128
　structure, 120

C

California white pine, *see* Yellow
　pine

White fir. Identification, 136
White oak. Identification, 126
  rules for grading, 158
  structure, 118
White pine. Structure, 123
  Eastern identification, 133
    rules for grading, 141
  Western identification, 133
  see also Yellow pine, Western
White spruce. Identification, 134
Width of pieces, 101
  joints, 43
  stock, 43
Willow. Structure, 120
Wire bands, *see* Metal binding
Wirebound box, 67
  assembling, 105
  closing, 107
  construction, 104
  grouping of woods, 104
  material, 104
  specifications, 103
  with detached tops, 107
  with wedgelock ends, 105
Wood fibers, 113
Woods used in boxes and crates.
  Amount of each species, 2
  availability, 1
  choice, 2
  cost, 1
  choice in relation to design, 45
  compression strength, 26
  density, 15
  description, 126
  desirable qualities, 2
  distribution, 5
  fiber saturation point, 16
  fungous growth, 31
  geographical distribution by
    States, 126
  grouping, 100, 104
  hardness, 30

identification, 115
kinds, 3
mechanical properties, 26
moisture content, 15
  see also Moisture content
nail holding power, 15, 30
nailing qualities, 51
odor, 26
physical properties, 14
properties which influence use
  in box construction, 14
resin content, 15
salvage value, 1
shearing strength, 26
shock resisting ability, 22, 30
sources, 5
species, 2
specific gravity, 14
stiffness, 30
strength properties, 26
structure, 108
taste, 26
tensile strength, 26
veneer, 38
weight, 14
  see also Lumber
Worm holes, 10
  · effect on box strength, 51

Y

Yellow buckeye. Identification, 132
Yellow pine, Southern. Identification, 134
  rules for grading, 143
  structure, 123
  Western. Identification, 134
    rules for grading, 148
    structure, 123
Yellow poplar. Identification, 131
  rules for grading, 150
  structure, 121

Made in the USA
Lexington, KY
23 July 2018